The Eugenic Mind Project

The Eugenic Mind Project

Robert A. Wilson

The MIT Press
Cambridge, Massachusetts
London, England

© 2018 Massachusetts Institute of Technology

All rights reserved. No part of this book may be reproduced in any form by any electronic or mechanical means (including photocopying, recording, or information storage and retrieval) without permission in writing from the publisher.

This book was set in ITC Stone Sans Std and ITC Stone Serif Std by Toppan Best-set Premedia Limited. Printed and bound in the United States of America.

Library of Congress Cataloging-in-Publication Data is available.

ISBN: 978-0-262-03720-4

10 9 8 7 6 5 4 3 2 1

For Leonie
and
in memory of Leilani
(1944–2016)

Contents

Preface and Acknowledgments

The Eugenic Mind Project derives from some unexpected twists and turns in my professional and personal life over the past ten years or so. Prior to that time, I had worked primarily in the philosophy of mind, cognitive science, and biology, and was turning to complete a projected trilogy of books on the individual and individualism in what I called "the fragile sciences." Having published *Boundaries of the Mind* on cognition in 2004, and *Genes and the Agents of Life* on biology in 2005, it was a matter of turning to the third book—on sociality. While I had had to absorb some new literature from the cognitive and biological sciences between 1998 and 2003 in writing *Boundaries* and *Genes*, moving to complete this trilogy on the role and conception of individuals in the cognitive, biological, and social sciences involved a more dedicated program of study. About cognition, I antecedently knew a reasonable amount, having completed a doctorate in the philosophy of mind and worked primarily in the cognitive sciences during the 1990s. About biology I knew passingly enough, chiefly from teaching the philosophy of biology from 1995 while at Queen's University, and then by benefiting from several programs for faculty aimed at promoting the depth of one's interdisciplinary scholarship, sponsored by my next erstwhile employer, the University of Illinois, Urbana-Champaign. Extending some ideas in *Boundaries* about individuals in the cognitive sciences into the biological sciences in *Genes* primarily relied on a good enough match between my own partial, growing bioknowledge and the inherently integrative (and forgiving) nature of the interdisciplinary field covering the history, philosophy, and social study of biology.

But my own ignorance ruled the day when it came to sociality and the social sciences. Here the learning curve was to be a fair bit steeper. In fact, it only gradually became clear over a number of years that the proposed third book in the trilogy, *Relative Beings*, could not adopt the smorgasbord approach that governed the second book—a little bit of this, a little bit of that. *Relative Beings* would focus instead on kinship as a primary structure governing individuals and the forms of prosociality in which they were immersed. But then eugenics, and a different kind of learning curve, came along shortly after moving to Edmonton, Alberta, in 2000.

The initial drafts of what have become the chapters in the first half of *The Eugenic Mind Project* were written while I was the principal investigator for the Living Archives on Eugenics in Western Canada project, a role that provided a wealth of experiences that inform the book. As will be clear from the introductory chapter, the collaborative work that emerged from this project, and particularly the contributions of eugenics survivors to that project, were the sine qua non for my own thinking here. I would especially like to thank Judy Lytton, Leilani Muir, Ken Nelson, Glenn George Sinclair, and Roy Skoreyko not only for their courage in sharing their experiences and lives in detail in a variety of settings, but also for their resilience and commitment to working together with community advocacy organizations, academics, and students on eugenics past and present. I am also grateful to the project's other team leaders—historian Erika Dyck, community advocate Nicola Fairbrother, disability studies scholar and activist Gregor Wolbring, and humanities computing specialist Natasha Nunn—and to the multitalented project manager, Moyra Lang, for their co-direction of the project over a five-year period.

I would also like to acknowledge the important role that students and former students played here, especially Jacalyn Ambler, Emma Chien, Luke Kersten, Bart Lenart, Colette Leung, Ben McMahen, Faun Rice, Aida Roige, Amy Samson, Josh St. Pierre, and Mark Workman. All contributed significant content to the project's ongoing website, EugenicsArchive.ca, a site that I have drawn on repeatedly in my writing and teaching.

These collaborations would not have been possible without the generous financial support of the Social Sciences and Humanities Research Council of Canada through their award of a Community-University Research Alliance grant for the project in 2010, as well as from our major university and community partners, particularly the universities of Alberta, Calgary, and Saskatchewan, and Neighborhood Bridges.

The book as a whole was drafted in January 2016 in Melbourne, Australia, and revised in light of reviewer and other peer feedback during the course of the remainder of that year. I would like to thank three anonymous reviewers for MIT Press for their distinctive, encouraging comments on what was, at that time, a manuscript without notes at about two-thirds its final length, and the philosophy editor at the press, Phil Laughlin, for his support for publication. I am indebted to Luke Kersten, Matthew J. Barker, Faun Rice, and Mary Horodyski for comments on penultimate versions of the chapters in part I, and to Michael Bérubé, Ken Bond, Maria Kronfeldner, Sandra Harding, Julie Maybee, Milton Reynolds, Susan Schweik, Alexandra Minna Stern, and Alison Wylie for more general guidance in shaping up the final version. More nascent forms of the material here were presented at a variety of venues, including in talks each year during Alberta Eugenics Awareness Week (2012–2014); at the biennial or annual meetings of Cheiron, the International Society for the History, Philosophy and Social Studies of Biology, the International Society for the History of Neuroscience, and the Pacific Division of the American Philosophical Association; at keynote addresses to the Australasian Association of Philosophy in Wollongong, the Atlantic Canadian Philosophical Association in Halifax, the Philosophy of the Life Sciences Network in Gut-Siggen, and the student-led Philosophy-History-Politics-Conference at Thompson Rivers University in Kamloops; and at invited talks at the Central European University in Budapest, the University of Vienna, Pennsylvania State University, Lewis & Clark College, Concordia University in Montreal, the University of Alberta, the University of Calgary, and the University of British Columbia, Okanagan, in Kelowna. I would like to thank my hosts in each case, and audiences for their interest and probing questions and comments.

I have drawn on joint work with Luke Kersten, Joshua St. Pierre, and Matthew J. Barker, and I am grateful for their permission to do so, as well as for the collaborations themselves. The material in chapter 7 draws on both "Eugenics and Disability," written with Josh, with sections 7.5–7.8 stemming from Josh's contributions most directly, and a paper written with Matt, "Well-Being, Disability, and Choosing Children." The material in chapter 3 draws on work undertaken with Luke Kersten. I am grateful to Garant Publishing for permission to use material from "Eugenics and Disability," which appears in *Rethinking Disability: World Perspectives in Culture and Society*, ed. Patrick Devlieger, Beatriz Miranda-Galarza, Steven E.

Brown, and Megan Strickfaden (Antwerp: Garant Publishing), 93–112. The vignettes from eugenics survivor stories that appear in section 10.8 have been reworked from my "The Role of Oral History in Surviving a Eugenic Past," which appears in *Beyond Testimony and Trauma: Oral History in the Aftermath of Mass Violence*, ed. Steven High (Vancouver: University of British Columbia Press, 2015), 119–138.

January 20, 2017

I Eugenic Activities: Probing Eugenics

1 Standpointing Eugenics

The importance of specific kinds of lived experience for understanding and explanation has been underscored within philosophy by proponents of standpoint epistemology and ethics. In the humanities and social sciences more generally, standpoint theory emphasizes the positive value that thinking about knowledge and value from certain kinds of marginalized standpoint can have. *The Eugenic Mind Project* is very much a reflection on eugenics and disability from such a standpoint. As such, it represents what I will call "standpoint eugenics," aiming to explore eugenics from the standpoint of those who are survivors of its history. Standpoint theory itself remains contentious within the humanities and social sciences, and whether this book is *really* standpoint eugenics, or just standpoint-ish eugenics, is something I shall return to in part III, "Eugenic Voices." There I will critically explore the contributions and limitations of standpoint theory as a framework for understanding the nexus between eugenics, disability, social inclusiveness, and human variation that is my focus. But the locational complexities that arise in such an exploration need not obscure the prima facie appeal of standpoint theory, and why the insights of standpoint theory provide a promising starting point for a distinctive perspective on what I call "The Eugenic Mind," which refers to a way of thinking that has often been viewed as merely a relic of the past.

Standpoint theory was developed by progressive political thinkers and social activists in the nineteenth century, and derives from a perhaps innocuous-sounding observation: that sometimes the people most adversely affected by a rule, policy, institution, practice, or entire system are in the best position to understand its nature, not only despite but also because of their adversity. This observation is close to home for many of us, and you don't have to travel into the flights of fancy of abstract philosophical and sociological theorizing to find examples. A child who is picked last (or not at all) to play in a friendly game of soccer or ice hockey because he is fat, or because she is a girl, feels the hurt that his or her peers cause through their choices. But he or she might also be able to sense that something isn't right here, whereas his or her peers remain oblivious; the child might even be able to tell you what it is like to be discriminated against in that way that passes beneath the radar of some of us, or that remains just a distant memory for others. A friend of mine once told me of a common experience he shared with many of his fellow African Americans: of being followed not so subtly around a department store by a security guard, constantly under watch and suspicion. The discomfort and wariness this kind of social imposition causes—like the hurt of being socially left behind because of your body shape or your sex—can grind you down, if it becomes regular and systematic, as it so often does. Yet such social impositions create both the human adversity and the space that I called the "best position" for understanding, a standpoint from which one can see or grasp more of what is really going on, a reality that remains masked from those in positions of relative advantage.

As one might expect, if only from the name itself, standpoint *theory* is something more complicated and convoluted than just this kind of observation, and we'll see what that "more" entails in chapters 9 and 10. Standpoint theory originates in theories of revolution and radical social change, particularly in Marxist views of the working class and capitalism, and has its most influential, recent expression in the feminist work of social scientists such as Dorothy Smith, Nancy Hartsock, and Patricia Hill Collins. As a theoretical contribution to the humanities and social sciences, standpoint theory is wrapped up deeply in talk of marginalization, systematic oppression, and epistemic privilege. But we don't need that talk to get some idea of what a standpoint *eugenics* might be. Standpoint eugenics rests on the idea that those who have felt the impact of eugenics—in the first instance, those who have lived through a eugenic past, and those confronting continuing eugenic thought and practice in their own day-to-day lives—in virtue of that lived experience face adversities and as a result have insights into eugenics that the rest of us lack, or gain at best second-hand. A standpoint eugenics is an account of eugenics from such standpoints.[1]

***

Eugenics arose in the nineteenth century both as a proposed, meliorative science—a science of human improvement or betterment—and as a social movement. Its central and most distinctive aim was to construct and use scientific knowledge and technology to regulate the sorts of people there would be in future generations, primarily by enhancing and restricting the reproduction of different sorts of people in the present generation. As animal and plant breeders had known for centuries, human agency can directly promote the reproduction of preferred specimens, as well as restrict or eliminate the reproduction of those deemed less desirable, in order to produce a population in future generations that has more desirable traits. Eugenics is based on the idea that the same is true when those specimens and individuals are members of *Homo sapiens*.

Suppose that you are growing corn in a small field with a few thousand plants. You notice that the corn plants vary in how rich the yellow of their kernels is, and that some plants seem less thirsty than others, requiring much less water to thrive. If of all the plants that are growing in your field you prefer those that yield bright yellow kernels, or that require less water, you can increase the presence of those traits in the next generation by mimicking what would happen in nature, were bright yellow and low thirstiness traits that differentially contributed, in some way, to the survival of those plants. You simply choose plants that have the preferred traits as your stock for the next generation, discarding those that produce dull yellow kernels and those that consume copious amounts of water. Repeated over generations, you might, with some genetic and environmental luck, even end up with a field of corn that is uniform with respect to these traits, a field of bright yellow corn requiring little water. This is selective breeding, and it was very familiar to plant and animal breeders in nineteenth-century Europe.

Familiar enough that it is no accident that Charles Darwin begins his most famous work—*On the Origin of Species*—with a chapter on selective breeding, what Darwin calls *artificial* selection, in laying the groundwork for the very idea of *natural* selection. In doing so, Darwin was using a traditional explanatory ploy, starting with the familiar as a way to explain the novel. For his new idea of natural selection was postulated as a mechanism that, like familiar artificial selection, could produce changes in populations over generational time. It simply did so without there being a selector or artificer, some person directing the process. It was selection in the absence of a creator.

As Darwin's first cousin, Francis Galton, knew from the outset, there was no reason why selective breeding could not be applied, at least in principle, to human populations. At least there was no *biological* reason. We were continuous with the rest of the living world, and just one particular species of animal among many. In addition to more detailed scientific knowledge about which traits were heritable (and how), and technologies that would allow us to measure traits and assess populations accurately, two other factors were needed to move eugenics from the glimmer of an idea to a series of social practices. The first was a sympathetic public, one that understood what good this knowledge would do for our species when it was translated into public policy. The second were governments prepared to introduce that public policy, policy motivated by their firm sense of what was right for the future of humanity. But Galton also knew early on that both the public and government would need convincing through the agency of science and the cultivation of public opinion. So he was poised to take eugenics down the dual pathways of science and social movement, even before he coined the term "eugenics" in 1883.[2]

***

Eugenics emerged in the late nineteenth century at the intersection of the then-nascent sciences of sociality—criminology, sociology, and anthropology in both its physical and cultural manifestations—and the emerging biomedical regulation of human populations through technological means. All of these fragile sciences of sociality involved, to some extent or other, new statistical measures and tools for the assessment of human abilities and capabilities. This involvement is sometimes taken to signal eugenics' construction of a notion of normality that systematically regimented what sorts of people there should be. I shall provide my own view of the confluences that led to the emergence of eugenics in chapter 2, returning to discuss in more detail the role of the concept of the normal in that emergence in chapters 5 and 6. Here I want to focus on something often glossed over in discussing the origins of eugenics: the idea of human betterment.

As a new science of human improvement, eugenics reflected a particular, prosocial tendency that runs deep in us as a species. That tendency drives so much of our social life that it is taken for granted in our everyday lives, sufficiently so that it can be barely noticeable. But it is there in thousands of micro-interactions we each have every day, interactions writ large in our systems of communication, in the cooperative conventions that create the norms by which we live, and in how we think of ourselves in terms of social identities and values that are structured by a kind of shared, collective intentionality. Human language, the social rules we have created to regulate how we should treat each other, and our sense of belonging, often deeply, to certain groups—our family, our nation, our people, our species—are all signs and products of our specific social nature. We are a species comprised of individuals who are distinctively socially integrated and socially interdependent, a species geared with tendencies that reflect our distinctiveness here.

Of the different ways to bring out the depth to these prosocial tendencies, I want to identify just one feature of our nature that brings it home. That is the extreme vulnerability that each one of us has from birth that makes our continued existence dependent on quite active forms of basic assistance and aid—food, shelter, protection, removal from harm's way. Each of us starts as a small, only partially formed creature that relies not only on not being neglected by others, but also on the often concentrated and costly investment that others have in us. This basic fact about our developmental reliance on the sociable goodwill of others is part of the reason that such sociability is a feature of our nature with sufficient commonality to ensure our survival to adulthood.

The fact that each of us moves from a state of vulnerability and dependence as we develop through the sociable good will of others tell us something important about human nature. We are not just sociable creatures in being socially related to one another in various ways, but *prosocial creatures*. Prosocial behavior flows from us, encompassing the helping of individuals, our volunteering of time, effort, and resources to benefit groups whose goals we endorse, and the taking on of sometimes even extreme risks and costs in engaging in altruistic behaviors, including risks that can cost us our lives.

We might keep in mind this prosocial nature as we acknowledge the eugenic moments of our collective past. Sometimes for better, sometimes for worse, the aim of human improvement is a manifestation of our prosocial nature. Yet the particular ways in which eugenics sought to bring about human improvement also, I want to suggest, reflect the limitations of this tendency, the ways in which our prosociality has been and continues to be circumscribed and bounded in practice.[3]

***

General academic interest in eugenics is relatively recent and has been largely historical in nature. Yet while much of the scholarship on eugenics has promoted the neutrality that historical distance often affords, the study of eugenics has seldom taken the form of distant, scholarly inquiry. Instead, that study typically has been undertaken with one eye on the present, with the history of eugenics conversely sometimes forming the scholarly backdrop for discussion of a diverse cluster of contemporary issues about science, technology, and human nature. Such issues include those concerning reproductive autonomy, science and scientism, biological (particularly genetic) determinism, and disability and human variation. This interplay between eugenics past and biotechnology present catalyzed during the 1980s, starting with the focus on the historical relationship between the biological sciences and eugenics in the historian of science Daniel Kevles's *In the Name of Eugenics*, and continuing with the exploration of genetic technologies and eugenics in the sociologist Troy Duster's *Backdoor to Eugenics*. As Kevles said more dramatically in his opening comments to his preface to an edition of *In the Name of Eugenics* published ten years after the book first appeared, the "specter of eugenics hovers over virtually all contemporary developments in human genetics, perhaps even more now than when this book was first published."

One form that specter took, and a focus of attention for discussions of such developments, centered on the Human Genome Project, the first big science project in the biological sciences and perhaps the largest collaborative endeavor in the history of science. It was initiated in the 1980s in a series of meetings, conferences, and workshops led by molecular biologists, biochemists, and geneticists, and funded to the tune of $3 billion by the U.S. Department of Energy and the National Institutes of Health. Officially beginning in 1990, the Human Genome Project's primary technical aim was to provide complete listing of the DNA sequencing found across the twenty-three pairs of chromosomes that make up the human genome. This listing is a series of chemical base pairs, guanine-cytosine (G-C) and adenine-thymine (A-T), whose discovery by Francis Crick and James Watson in 1953 as the fundamental building blocks in the double-helix structure within human chromosomes was the basis for the Nobel

Prize for Physiology or Medicine that they shared with the physicist Maurice Wilkins in 1962.

But the Human Genome Project's relevance for those interested in the resonances of eugenics past lies in the perceived significance of that listing and the corresponding motivations for the project itself, along with the assumptions made linking biotechnological achievement with those motivations. For high among those motivations were variants on classic eugenic talk of human betterment, cast both in terms of enhanced understanding and improvement of human nature and in terms of addressing the effects of human illness and disease through elimination and cure. As many had made clear from the outset, linking basic, molecular knowledge of the kind provided by the Human Genome Project to higher-level claims—about genes, about traits, about therapies—required much knowledge that was certainly neither biochemical nor molecular. More generally, extending such linkage through to practices and policies also required value judgments—about disease, disability, and human variation, for example— that were not purely scientific matters. There were more than shades of a eugenic past here, where the gap between scientific enthusiasm and potentially socially impactful, devastating outcomes was bridged by controversial and fatally mistaken judgments about the lives of many of our most vulnerable citizens.

General wariness about the Human Genome Project was heightened by the appointment of James Watson as the director of the Office of Human Genome Research in 1988, a position he left in 1992 following disputes over patenting and alleged conflicts of interest. Watson had been the director of Cold Spring Harbor Laboratory since 1968, a site of special significance in the history of American eugenics (as we'll see further in chapter 2). Although Cold Spring's involvement in eugenics ended before the conclusion of the Second World War, Watson was known to express eugenic-sounding views, which became more pronounced and he expressed more frequently in his life after leaving the Human Genome Project. In general terms, Watson has long made it clear that he holds genetic knowledge to be key to understanding differences between human social groups (such as races) and particular human traits (such as homosexuality). Equally clearly, Waston views genetic technologies, including those enhanced and advanced through the Human Genome Project, as important ways both to promote desirable traits and to remove undesirable traits from the population.

More recently, the contemporary resonances of the eugenic past have spiked, with the ways in which eugenics has been taken up by both the academic community and the public, diversifying and becoming increasingly connected with issues of ongoing significance for people marginalized in our societies by eugenic ideas, practices, and policies. For example, in recent years we have seen the publication of a major handbook on the history of eugenics, several journals that have dedicated special issues to eugenics, books exploring eugenics in North America in more detail, as well as those focused particularly on eugenics in Alberta, and the appearance of eugenics survivor testimony and memoirs.

The Eugenic Mind Project is a part of this more recent turn in the study of eugenics, a turn that brings eugenics home, both from a perceived distant past to the ongoing present, and from ideas and practices that primarily affect others to those that remain continuing issues for many of us in our daily lives. Eugenics has become increasingly a subject for scholars from a variety of disciplines who draw on resources made possible by the testimony and reflection of those lacking the comfort of arm's-length distance from a eugenic past.

So the social relevance of the study of eugenics stems not simply from the long shadow cast by the genocidal, government-directed policies of the Nazis in the name of eugenics, but also from the permanent possibility of eugenics gaining purchase through the back door of technologically-enhanced individual choice. But recognition of the need for public engagement and activism around the topic of eugenics is, I think, better understood against two dissonant recent social contexts that have been especially poignant in North America.

First, in the early 2000s, there were official apologies from the governments of four of the thirty-three American states to have passed eugenic sterilization laws: Virginia, Oregon, and North Carolina in 2002 and California in 2003. These apologies themselves followed in the wake of settled legal actions in Alberta, Canada, brought by eugenics survivors against the government of Alberta for wrongful confinement and sterilization under the province's Sexual Sterilization Act (1928–1972), and at least some of those advocating for those apologies were keenly aware of the situation across the northern border. In addition, during this time there was a growing interest in North America in recent revelations of enduring eugenic sterilization in the Scandinavian countries of Sweden, Norway, Denmark, and Finland.

Second, though the settlements and apologies aimed to make it clear that eugenics was a matter of a regrettable past, that view of eugenics has seemed complacent to many within the disability community who see eugenic thinking manifested in a variety of contemporary practices. A prominently discussed example is the widespread practice of selective abortion that draws on disability-focused genetic screening and testing, such as that for trisomy 21 or Down syndrome. Here it is not abortion per se but *selective* abortion aiming to prevent the birth of people who would likely manifest specific disabilities at some point in their life that has been the focus of critical attention. For—as the objection is sometimes rhetorically put—if we hold that selective abortion on the basis of sex or gender (e.g., being female) is at best morally problematic, how can that practice be justified when the discriminating characteristic is not sex but disability? This disability rights critique of selective abortion has opened up questions about The Eugenic Mind concerning reproductive rights, questions that will be the focus of discussion in chapter 7.

More striking, however, is that the view of eugenics as simply a regrettable past was undermined by recent revelations of the ongoing sterilization—in Australia, in California, and in India—of just the sorts of people who were the target of past eugenic policies and laws. I shall return to briefly discuss these cases in the conclusion of this chapter. But the simple reminder that even old-style eugenics is not only a historical matter makes all the more pressing questions about the forms of eugenic policy beyond sterilization, and about the manifestations of "newgenic" thought and practice that exist now.[4]

***

I had become aware of the not-so-distant history of eugenics in the Western Canadian province of Alberta shortly after arriving in Edmonton to take up a position as a professor of philosophy at the University of Alberta in July 2000. By April 2004, I had met a range of people involved in the case commonly referred to as *Muir v. Alberta* and, more long-windedly, *Muir v. The Queen in Right of Alberta*, a legal decision that I had not heard of prior to my arrival in Alberta. In this landmark legal case decided in 1996 by Madame Justice Joanne Veit, eugenics survivor Leilani Muir successfully sued the province of Alberta for wrongful confinement and sterilization relating to her admission to and treatment at the Provincial Training School for Mental Defectives in Red Deer, Alberta, from 1955 until 1965. As a child of ten, Leilani had found herself swept up by the eugenics movement. After being institutionalized, Leilani was sterilized putatively in accord with the Sexual Sterilization Act of Alberta, a law that was in place in the province until 1972. That provincial law, one of only two enacted in Canada's history, authorized the eugenic sterilization of individuals whose recommendation for sterilization by the medical superintendents of provincial institutions or other state authority figures had been approved by a four-person committee known informally as the "Eugenics Board." The legal wrongfulness of both Leilani's institutionalization/confinement and her sterilization that was established in *Muir v. Alberta* drew attention to many problematic features of how eugenics was practiced in the province, including how the Eugenics Board did its work.

Alberta's western neighbor, British Columbia, was the only other province in Canada to implement sexual sterilization legislation, with the British Columbian legislation modeled on that in Alberta but enforced much less rigorously. This is not to say that eugenics existed only in these two Western-most provinces; Saskatchewan and Manitoba, the other two Western Canadian provinces, came very close to passing eugenic sterilization legislation in the early 1930s. In fact, sterilization legislation passed in Saskatchewan in 1930 but the government fell before it could be put into practice; in Manitoba, the 1933 Mental Deficiency Act that included provision for eugenic sterilization was just one vote away from becoming legislation. In addition, much of the self-sustaining momentum for eugenics in Alberta was initially generated by the advocacy of influential figures in the Canadian mental hygiene movement based in the province of Ontario, such as Helen MacMurchy, C. K. Clarke, and Clarence Hicks. There is also much evidence of the influence of the reach of eugenics in Canadian social policy beyond explicit eugenic sexual sterilization legislation.

Among those I met connected to the Muir case were people called as expert witnesses in that and subsequent cases, lawyers involved directly in those cases, and people with general academic expertise in eugenics: from law, sociology, history, education, genetics, psychology. Many of them were my campus colleagues, housed in other departments and faculties on campus. But most important, I met Leilani Muir herself, a meeting that, little did I know at the time, would change much of my work and personal life, and shape my perspective on eugenics and disability.

The briefest way to pinpoint why, at least in terms of how Leilani struck me initially, was that Leilani, then in her late fifties, was not noticeably different from other people I knew. Leilani was distinctive, and admirably so as I got to know her better, but not *different* in the way one might expect, given her history. She was, to put it in terms of a concept that structures our perceptions of human variation, about as normal as any of us can be. Yet Leilani had been institutionalized at a school *for mental defectives* for an extended period of time as a child, as a teenager, and as a young adult; she had also been classified as a "moron"—a term whose colloquial familiarity now might make it surprising to some to learn that it was invented barely 100 years ago by the eugenicist and psychologist Henry Goddard to pick out "higher grade" mental defectives. Classified as a higher-grade mental defective, Leilani was sterilized putatively in accord with the Sexual Sterilization Act of Alberta. And all of that had further, unexpected, and devastating consequences for Leilani's post-institutional life.

How did this happen? Leilani was certainly different from the educated, upwardly mobile, middle-class people who populated my snug university surroundings. But she wasn't that different from the less educated, often class-stagnant, working-class people with whom I grew up. And, it turned out, she was not that different from many hundreds, if not thousands, of others who were subjected to the very same laws and policies in Alberta. *How does this kind of thing happen?*

Yet this perhaps makes my side of the equation here sound more academic than it in fact was. For this was not really a kind of intellectual conundrum to be solved, one that, as a card-carrying analytic philosopher of mind and science, I was no doubt overprepared to solve. This was not a puzzle case, some real-life analogue to the thought experiments that, for better or worse, still drive so much of the best contemporary work in analytic philosophy. It was much more personal and emotional than that, for at least two reasons. Those reasons correspond to two dimensions to eugenics as practiced that continue to function for me as constraining anchors for reflections about eugenics past, present, and future. These were issues of what I will call *institutional complicity* and *engaged individuality*.[5]

***

First, institutional complicity. Paramount here for me was that my own university—indeed, my own department—had been centrally implicated in the history of eugenic sterilization in Alberta. This was primarily through the agency of the founding chair of my department, Professor John MacEachran, who also served as the original head of that committee known as the Eugenics Board. And MacEachran had done so from its founding in 1928 right through until 1965, meaning, for almost its entire history. As one of the original professors hired early in the history of the University of Alberta, MacEachran quickly rose in 1914 to become the first provost at the university, and continued to serve as the chair of its Department of Philosophy and Psychology as it became the Department of Philosophy, Psychology, and Education in 1933. In fact, MacEachran continued in these roles, as department chair and provost, until his retirement in 1945. He was, and remains, the longest-serving department chair, and the longest-serving provost, in the 100-plus-year history of the University of Alberta.

Provost MacEachran was a fellow philosopher by training, having completed both a doctorate in the subject of philosophy at Queen's University in Canada, and another at the University of Leipzig in Germany. There is apparently (and strangely) no record of his doctoral dissertation at Queen's University, not even its title, except records indicating that his doctorate was taken in philosophy, and his Leipzig dissertation is a short treatise, in German, on pragmatism and knowledge. Despite his limited education in the psychological and biological sciences and without a single research publication in those highly relevant areas, MacEachran had, for more than thirty-five years, chaired a board that approved approximately 4,800 eugenic sterilizations. He served in this capacity for twenty years beyond his formal retirement from the University of Alberta until just seven years before the Sexual Sterilization Act of Alberta, which had created the Eugenics Board in 1928, was repealed in 1972. MacEachran was still serving as the chair of the Eugenics Board, giving approvals to applications for sexual sterilizations, when he was eighty-eight years old. However, in the wake of *Muir v. Alberta*, it had become clear, all too late, that the Eugenics Board under MacEachran's leadership had failed spectacularly to uphold its basic function to ensure that sterilization was imposed only on those who strictly met the conditions of the Alberta statute.

While the distinctiveness of the Sexual Sterilization Act is something I will return to in chapter 3, here we can note that it set two basic requirements to be met for sexual sterilization to be approved. The first was that release from the institution be imminent (in cases where the person recommended for sterilization was institutionalized); the second was that there was evidence that the mental deficiency of the "inmate" or "patient" would be transmitted to any offspring. Yet the Muir case showed that sterilization approval was often given by the Eugenics Board when release from the institution was not imminent, and frequently there was little evidence for the heritability of any mental deficiency. In fact, often there was much counter-evidence, which was in effect ignored by the board. Although the board neglected its basic duties in various other ways, the significance of its failure here and its consequences for Leilani were made clear in the decision of Madame Justice Veit, whose ruling included the summarizing statement that the "circumstances of Ms. Muir's sterilization were so high-handed and contemptuous of the statutory authority to effect sterilization and were undertaken in an atmosphere that so little respected the plaintiff's dignity that the community's and the court's sense of decency is offended."

In the course of the development of the legal case, it also became apparent that Leilani was far from alone in her legally wrongful confinement and sterilization. Leilani had not simply been one child who, through misfortune, carelessness, or administrative neglect of duty, had ended up in an institution that she should never have been admitted to, and sterilized under the authority of a law whose rubric she did not fit. Leilani was one of many. Very many. The failures here, in other words, were systematic and widespread. The ramifications had been devastating for large numbers of children confined and sterilized in Alberta between 1928 and 1972, whether wrongfully or not.

Leilani's case, including the important records made public by Madame Justice Veit in her decision, motivated more than nine hundred subsequent legal actions. The vast majority of these actions were eventually settled by the province of Alberta for over $80 million. Those settlements ended several years of attempts by the government of Premier Ralph Klein to dismiss the legal actions. The most controversial at the time, but perhaps least well known now, of the twists and turns to these attempts was the government's introduction of Bill 26 to the floor of the provincial legislature in March of 1998. Bill 26 appealed to the clause of the Canadian Charter of Rights and Freedoms commonly known as the "Notwithstanding

Clause" and sought to limit the amount of any settlement made to eugenics survivors to $150,000, and to do so independently of what was found to have happened to those confined and sterilized. Until that time, the Notwithstanding Clause was designed for, and had occasionally been used by, provincial governments to exempt themselves from certain Charter requirements that would impose an undue burden on them and the people they represented, given special circumstances that a province might find itself in. It had rarely been appealed to previously, and when it had, it was typically associated with the distinctive linguistic and cultural circumstances of the predominantly French-speaking province of Quebec. The unprecedented move of the government of Alberta invoking that clause to limit its settlement payments to sterilization survivors—who had allegedly been wrongfully confined and sterilized as part of past government eugenic policy—triggered an avalanche of criticism of the government. This public backlash was sufficiently strong that, even in a province in which the government held an overwhelming majority of seats, and had held power since 1972, the legislation was withdrawn within twenty-four hours. Following that, the government relatively quickly settled hundreds of remaining cases. Eugenics then became, once again in Alberta at least, a matter of the past.[6]

***

Further from the public eye, Leilani's case became central to the work of a small departmental committee, formed in 1997 and tucked away in the Humanities Centre at the University of Alberta. This three-person committee, informally known as the MacEachran Sub-Committee, within the Department of Philosophy at the university issued its *MacEachran Report* in 1998. This report was endorsed by the department as a whole and summarized many of the key failings of the Alberta Eugenics Board as they had been detailed in the public records of the Muir case. It also recommended courses of action, such as the removal of MacEachran's name from university prizes and other forms of honor, as well as advocating the teaching of this aspect of the history of eugenics within regular courses in philosophy and at the University of Alberta more generally.

MacEachran was not, of course, the only person facilitating eugenic practices courtesy of his or her position at the University of Alberta. Among the others, prominent was Professor Margaret Thompson, the first person qualified as a geneticist to briefly serve on the Alberta Eugenics

Board in the early 1960s while she was an assistant professor. (Thompson's own role in *Muir v. Alberta*, testifying as a witness for the province, played an unexpectedly significant and chilling role in that case; more on this to follow.) But MacEachran's involvement brought matters very close to home for me. MacEachran completed his M.A. and (reportedly) his first Ph.D. at Queen's University, where I held my first academic position—again, in the very same department—and he had briefly studied with one of the founders of the discipline of psychology, Wilhelm Wundt, at Leipzig, in the first decade of the twentieth century. I had recently written a little on Wundt in talking about the characterization of individuals in early psychology in *Boundaries of the Mind*, alongside an equally brief discussion of what I called "Galtonian individuals," taking up how one of the founding figures in the history of eugenics had thought of the nature of individuals in psychological science. And I had recently been hired at Alberta as a professor with an expectation of working together with others to bridge the separate departments of philosophy and psychology as part of an exploratory pathway to bringing cognitive science proper to the University of Alberta. Before its bifurcation in 1963, the department had been called the "Department of Philosophy and Psychology," including for nearly all the years during which MacEachran served as the chair of the Eugenics Board.[7]

***

So the issue of institutional complicity was, professionally speaking, close to home. The second more personal and emotional dimension to my reaction concerned Leilani. Being normal—whatever that is exactly—is just the beginning of how Leilani struck one immediately upon meeting her, an appearance shaped no doubt by one's expectation that there must be *something wrong with her*: Leilani was, after all, institutionalized for ten years at a government-run training school for "mental defectives." In fact, to this day, when others I meet who know about Leilani second-hand and then learn of my close friendship with her, the question they most often ask me is "What was wrong with her?" or its whispered variant, "What did she have?"

Yet Leilani sparkled with her own individuality and idiosyncrasies. I got to see more of these the more time we spent together, especially as we drifted further away from talking about eugenics and its destructive role in her life. Animals were a deep love of Leilani's. As were children. And Elvis! Generosity and well-meaningness spilled out from her. Having to eat second and third helpings of pretty much everything that Leilani had spent the previous day preparing for a "little get together" at her house—not that I would object *that* much—was pretty much required. It was something that brought me back to memories of two nanas, my maternal and paternal grandmothers, each of whom, in their own way, had almost the same degree of determined insistence to make sure that no one left their house with a rumbly tummy.

That engaged individuality, as I would come to think of it, involved distinctive personality traits, interests, likes and dislikes, self-concern (both short- and long-term), a kind of caring, social concern and feeling for others, a sense of one's own agency and its limits, and an imaginative ability to connect self and other. It was something that I found, in some way or other, in the hearts and minds of all of the relatively small number of sterilization survivors who I came to know well over the coming years, even if I felt it most strongly in interacting with Leilani. Many of those institutionalized and sterilized as "mental defectives" in Alberta over a forty-year period had manifested the characteristics I am thinking of as forms of engaged individuality. They had the aspirations, goals, and sense of themselves as agents that are the marks of individuality, and each of them was very much his or her own person. And they engaged not simply solipsistically with their own mental worlds but also with other people, striving for their own version of a sense of community and belonging. This was certainly true of other eugenics survivors I came to work with and befriend, such as Judy Lytton and Glenn George Sinclair. What did their engaged individuality imply about the practice of eugenics in Western Canada, and perhaps more generally?

The engaged individuality that one could readily feel flowing from survivors of eugenic institutionalization and sterilization—once they were given the opportunity to share their stories in a safe and welcoming space—contrasted sharply with what one could find in the expressed views of leaders of the eugenics movement, including those in the Alberta eugenics movement. This was especially so when it came to their characterizations of "mental defectives," the feeble-minded menace to society, and the subsequent life-devastating interventions of eugenic enthusiasts. Those interventions were sometimes very close at hand. They typically indelibly marked the lives of people—many people—who were subhumanized as Leilani had been as a ten-year-old child.

There is little reason to think that such leaders of our society—local, provincial, and federal politicians; university presidents, provosts, deans, and college professors; government officials, community organizers and advocates, and journalists—lacked engaged individuality in other aspects of their own lives, even if for some of them their personae were no doubt dulled by the demands of a professional and public career. They displayed much care and concern for their loved ones, and typically saw themselves as acting for the good of their community; they were as prosocial, cooperative, and humane as anybody else. So how was it that they could display such a glaring and presumably isolated deficit in this crucial dimension of humanity when it came to those deemed—often recklessly—unfit to parent, biologically and socially? What failures of humanity—of imagination, of concern, of agency—were in play here?

To be clear, for me this glaring deficit is as much a matter of substance as it is of expressive tone. In other words, it is not simply a matter of offensive or insensitive language. Rather, it is also a matter of the views themselves, views that, in effect, subhumanize people thought to be of a certain sort or type.[8]

***

These twin dimensions to eugenics as practiced—deep institutional complicity (and more than complicity), and a cluster of questions about engaged individuality—began for me as matters of the past. But over time, I came to see them as more than that: they were very much matters of ongoing attitudes, practices, and policies. And they were more closely related than one might initially suspect.

By the late 2000s, my perspective on eugenics had moved through a series of changes. I began from the supposedly objective perspective of a philosopher of mind and biology interested in eugenics as a case study in the relationship between science and values. Within a few years, my perspective had shifted to that of an ally of survivors of a long-lasting eugenics program discovered in what had, in the meantime, become my very own backyard. In the process, I had formed professional relationships with academics in other fields, such as sociology, history, and law, who knew much more about eugenics in general and in Alberta than I did. And bridges were built across what I had learned to call the "town/gown divide" back when I was a graduate student in upstate New York.

This community building was at the heart of what became the Living Archives on Eugenics in Western Canada project, funded generously by the now-superseded (some might say "abandoned") Community-University Research Alliance (CURA) program of the Social Sciences and Humanities Research Council of Canada. CURA had sought to encourage those in the humanities and social sciences to develop programs of research that both drew on and contributed to life beyond the academy, providing five years of funding for successful projects, and leveraging substantial community-university partnerships.

The Living Archives project aimed to examine the eugenic past in Western Canada and its contemporary significance by placing the engaged individuality of eugenics survivors at the heart of the project. Those who had lived through Alberta's eugenic past were not so much "human subjects" of oral history work as key co-participants to building a range of public resources for exploring eugenics, past and present. Some served on the project's governing board. Together with other survivors they participated in the public events we sponsored as part of Alberta Eugenics Awareness Week each October and in our summer intern program for students. All of them exercised ultimate control over the content and style of the video stories that form a crucial part of those resources.

Perhaps the most important other thing to say about "the CURA"— our own unimaginative, shorthand way of referring to the Living Archives project—is that of all the community partners who might have had (and in many cases, did have) the opportunity to participate in a long-term project focused on the history of eugenics in Western Canada and its ongoing significance, the most enduring partnership that was formed was with a small, local, insecurely-funded disability advocacy group, Neighborhood Bridges. Neighborhood Bridges saw immediately, and lastingly, the importance of the local history of eugenics for the lives of those with a variety of disabilities, including especially intellectual disabilities. They did so as they struggled to find and secure acceptable housing for people with disabilities who were battling for a modicum of independent living beyond institutionalization and group homes, and fought for their rights to parenting and intimacy. In short, both Neighborhood Bridges and those they represented and stood with were positioned to adopt a eugenic standpoint, a point of view that made the history of eugenics very much their own history. This was so despite the fact that, in virtue of their age and the end of the explicit eugenic era in Alberta in 1972, very few of those in the Neighborhood Bridges community had the lived experience of the kind that Leilani, Judy, Glenn, and others had had in growing up in the Provincial Training School for Mental Defectives. Yet they were eugenics survivors of a kind, and their involvement in the Living Archives project was to significantly influence what the CURA became, as well as my own perspective on eugenics past, eugenics future.

So the unavoidable ethnographic intrusion into the ongoing lives of eugenics survivors was, for us, in part a form of community building. This community building aimed not simply to compensate for the limitations to community that derived from a history of institutional isolation and stigmatization, but also to create a sense of substantial community among a more diverse range of people who thought of themselves as survivors of eugenics in a broader sense. That community was multiply intergenerational, fueled by the two-way identification between older eugenics survivors—like Leilani, Judy, and Glenn—and younger folks engaged by, and even living their lives shaped around, disability and the struggles they confronted in their day-to-day lives.[9]

***

Correspondingly, within the Living Archives project we came to think of eugenics survivorship in two ways: in terms of those whose lives had been governed fairly directly by laws such as the Sexual Sterilization Act of Alberta, and in terms of those people with disabilities in our local community who saw and felt in their day-to-day lives very much the same kinds of subhumanization and social exclusion that had been implemented through those laws and policies of years past. Among the latter were people with disabilities who were parenting, or were considering parenting options, in a broader community and culture with a certain history. That community had, not all that long ago, been led by people who had not only advocated for the sterilization of people "like them" to prevent the transmission of their putative defects to progeny, but had also deemed such people to be "incapable of intelligent parenting," to take a phrase that one can find on key forms that make up the administrivia of what I shall call *the social mechanics of eugenics* in Alberta.

Eugenics' social mechanics clearly needed attention if we were to grapple collectively and jointly with the eugenics survivorship of those with lived experience of a eugenic past. But such a social mechanics was also important for others *parenting around disability*, as we came to think of it. This included parents of infants and children with disabilities, particularly intellectual disabilities, who also came to see their own circumstances very much in light of the recent eugenic past in Alberta.

As the CURA developed over the next few years, however, we discovered what might be thought of as another kind of eugenics survivorship. Unlike the other forms this took, it was not locally based. Yet it came to play a pivotal role in our conception of the importance of what the Living Archives project was attempting to do. Eugenics survivorship here, in some sense, combined both of the previous forms we had encountered. Eugenic *sterilization* was, it turned out, not simply a matter of the past.

In 2012, the Senate of Australia—my beloved home country—launched an inquiry into the ongoing, often nonconsensual sterilization of girls and women with disabilities, a practice that had been brought to light through Medicare billing records. Unlike Canada and the United States, Australia had never passed sexual sterilization legislation, but the affinity between what was happening then and there in Australia and the broader eugenic past was part of what garnered the attention of the Senate. Floating free of explicit state-sanctioned policy, the documented practice of sterilizing women and girls with disabilities "for their own good" nonetheless often rested on eugenic arguments and, in any case, sat uneasily with Australia's formal human rights commitments, as argued in a detailed submission to the Senate inquiry by the advocacy group Women With Disabilities Australia.

During the summer of 2013, Cory Johnson of the Center for Investigative Reporting revealed that women in the California prison system had been recently sterilized under conditions of dubious consent or where consent was missing altogether. Johnson's reporting revealed that about one hundred and fifty Latina and African-American women had been sterilized between the years of 2006 and 2010, and the matter was put before the California legislature for discussion. As the state in which more sterilizations had been carried out than in any other American jurisdiction in the heyday of eugenics—about one-third of the then-legal eugenic sterilizations performed in the United States between 1907 and 1977 had occurred in California—legislators in the state were already very much aware of the need to acknowledge the legacy of a eugenic past. In the early 2000s, Governor Gray Davis's formal apology for California's eugenic history, together with California's Senate Resolution No. 20, had expressed "profound regret" over the state's involvement in eugenics. The Senate resolution had urged "every citizen of the state to become familiar with the history of the eugenics movement, in the hope that a more educated and tolerant populace will reject any similar abhorrent pseudoscientific movement should it arise in the future." Neither the apology nor this resolution were accompanied, however, by any meaningful public policy change, such as compensation or educational reform. In the wake of the 2013 Johnson report of ongoing sterilizations, what seems needed is not so much an acknowledgment of a eugenic past, but immediate steps to halt an ongoing eugenic present.

Finally, at the end of 2014, more than a dozen women in the central Indian state of Chhattisgarh died after undergoing sexual sterilization as part of a paid incentive program intended in part to control poverty through population containment. The women died of blood poisoning or hemorrhagic shock following sterilization, and the news story spread worldwide because few outside of India, and perhaps within the country as well, knew of the extensiveness and routine nature of this sterilization program. According to United Nations statistics compiled in 2006, as many as 37 percent of Indian women have undergone sexual sterilization, many as part of this incentive program, which offers free sterilization for women and pays them $10–$20, amounting to more than a week's salary for many of them.

These were far from isolated outbreaks of practices reminiscent of old-style eugenics. A few years earlier, the government of President Fujimoro had approved the use of sexual sterilization to curtail the indigenous population in Peru, resulting in perhaps three hundred thousand sterilizations, and there were continuing reports of Romani women in countries from the former Eastern Bloc being sexually sterilized without consent. In late 2015 and early 2016, Canada's national network, CBC, issued several reports detailing cases in which First Nations women had recently been sterilized without, or with dubious, consent in Alberta's neighboring province of Saskatchewan.

As I returned to make final revisions to this chapter, the Peruvian public prosecutor responsible for investigating charges of crimes against humanity levelled at President Fujimoro for his role in the sterilization policy had decided not to pursue those charges. The prosecutor is reported as saying that the practice was not a part of state policy, but rather was a series of isolated cases—a judgment formed despite over two thousand testimonials, many of which directly contradict this claim. Whether this view of the institutional complicity of the government of Peru in eugenic sterilization is itself a form of institutional complicity in The Eugenic Mind, I leave as an exercise for readers with engaged individuality.[10]

***

All of this raises, for me at least, questions that go to the heart of eugenics. Some are abstract questions about what eugenics is and what a standpoint eugenics amounts to; others are closer-to-the-bone questions concerning how to meaningfully commemorate eugenic history and acknowledge what continues to happen in the name of eugenics. These are all largely questions about the relationships between our eugenic past and what might be called its legacy today. Except that "legacy" at least softly suggests the after-effects of something that has moved through the waters of history, leaving just the ripple of its passing.

So perhaps *The Eugenic Mind Project* isn't really concerned with the "legacy" of eugenics. It aims to understand The Eugenic Mind—the nature of eugenic thinking, past and present. It does so, however, neither by conceptualizing that past as simply containing lessons for the present, nor by calling for projection of oneself into the minds of eugenicists themselves. Rather, it manifests a kind of standpoint eugenics, taking the perspectives of those who became, and those who remain, the targets of eugenic thought as key to understanding The Eugenic Mind. Within standpoint eugenics, The Eugenic Mind represents a way of thinking that is very much with us, as we will see.

2 Characterizing Eugenics

2.1 The Short History of Eugenics

The psychologist Hermann Ebbinghaus's famous quip that "psychology has a long past, but only a short history" has become a cliché for historians of psychology, as well as those thinking more generally about the disciplining of the mind through the development of the psychological sciences in the last third of the nineteenth century. What it states about the sciences of the mind—long past, short history—however, is also worth acknowledging in a variety of other contexts in the growth of theory and knowledge, especially those in which knowledge and technology coalesce in the formation of some new form of inquiry. Ebbinghaus's point is even more relevant in those contexts in which we find the kind of revolutionary fervor that drove the founders of psychology as a distinctive discipline, such as Ebbinghaus himself, and generated a kind of enthusiasm for the importance of the benefits for society of the corresponding emergent paradigms, methodologies, theories, and results.

The short history of eugenics is typically regarded as occupying the eighty years between 1865 and 1945, and might be thought to be a natural focus for *The Eugenic Mind Project*. The adaptation of Ebbinghaus's quip to the context of eugenics both serves to remind those focused on this short history, however, that there might be a longer past to explore and also raises the question of whether that short history is as short as is often supposed. Is eugenics a matter of our *history*—albeit a history that itself may inform our views of current ideas, practices, and policies? Or did eugenics go underground in the remainder of the twentieth century, continuing to surface as what some have thought of in new forms—*newgenics*—over the past seventy years? Can one find The Eugenic Mind now not only explicitly in the squalid quarters of the ignorant, the xenophobic, and the extreme, but also more implicitly in mainstream contemporary thought and social practice?

These are questions whose answers are developed in the book as a whole, and I shall return in the chapters in part II to discuss newgenics more directly in light of those answers. But I want to concentrate in this chapter on what eugenics is, beginning with the origins of its short history.

Eugenics arose, as did the sciences of the mind, in the second half of the nineteenth century promising a new kind of knowledge that was informed by the emerging biological, psychological, and social sciences— *the fragile sciences*. These fragile sciences of human nature could be put to use to make a positive difference in the quality of the lives of present and future people. The emerging scientific knowledge and technology were to shape what sorts of people there would and should be. The envelopment of eugenics within the biological, psychological, and social sciences was many-threaded. And the early scientific credibility of eugenics rested at least as much on appeals to the emerging psychological and social sciences as on the still nascent science of genetics and the biological sciences more generally.

The two-way connection between artificial and natural selection in Charles Darwin's work and the integration of *Homo sapiens* into the developing evolutionary paradigm, briefly mentioned in chapter 1, gave eugenics and the idea of eugenic selection a tentative grounding in biological explanation. That grounding was sufficient to at least raise the question, asked by Francis Galton (among others), of what barriers there are to applying what we know to be true of plants and animals to our own species. Yet this reliance of the nineteenth-century origins of eugenics on evolutionary biology and the biological sciences more generally should not be overstated. It is easy to forget from the genocentric perspective that directs both current scientific and popular culture that the science of genetics itself didn't crystallize until the early years of the twentieth century, and until at least that time the reception of the application of Darwinian thinking to the human world vacillated between the extremes of uncritical enthusiasm and resolute rejection.

Rather than being some kind of natural outgrowth of the biological sciences, the scientific credibility of eugenics was distributed in a more motley fashion across the fragile sciences. If anything, that credibility rested more squarely on the adolescent shoulders of the psychological and the social sciences. Eugenics drew on the measurement and classification of people in virtue of their mental powers, deficiencies, and dispositions in psychology; on ideas about higher and lower races of human species and the progression from savagery to civilization in anthropology; and on the detection of natural-born criminals and the diagnosis of irremediable social waywardness in criminology and sociology. This spread of both the origins and credibility of eugenics across the fragile sciences created the conditions for the position that eugenics came to occupy between applied, meliorative science and transformative social movement.

As a social movement, eugenics owed much to a utopian view of how these new fragile sciences could inform a large swath of social policies, including those governing health, reproduction, and immigration. While it has been tempting to some to view eugenics as moving from a mere idea in the last third of the nineteenth century to become an influential, scientifically informed social movement in the first half of the twentieth century, one that directly affected the lives (and deaths) of millions of people in all six populated continents, there was never a time when eugenics was simply an idea, or simply pure science that later came to be applied, institutionalized, and disciplined. Eugenics was articulated in relationship to ideas and sciences that themselves were forged in response to perceived social ills: problems of mental deficiency, of criminality, of the degeneracy of the human race. Eugenics carried with it an enthusiasm, both popular and scientific, for solving a wide-ranging series of social problems once and for all by controlling who reproduced, and who was reproduced, in future generations. That enthusiasm can be found at the very beginning of the short history of eugenics, in the life and work of Francis Galton.[1]

2.2 A Galtonian Start

Francis Galton (1822–1911) was what is sometimes called a "gentleman scientist" whose polymathic interests and skills were manifest throughout much of his long life. Galton studied medicine in London briefly before turning to mathematics at Trinity College, Cambridge. Having inherited sufficient funds upon the death of his father in 1844 to enable him to live without employment for the rest of his life, Galton was able to follow his interests wherever they led—from exploration in the Middle East and southern Africa to the development of statistical techniques for understanding population tendencies and the introduction of composite portraiture and fingerprinting as more specific methodological innovations designed for social purposes.

Galton's most encompassing passion was for the uses of science and technology to improve society and the people that constitute it. The most enduring influence on that passion was Galton's understanding of the relevance of Darwin's theory of evolution for human improvement. For Galton, it was not only that the theory of evolution held for *Homo sapiens*, as it did for any animal species. Rather, Galton was driven by his belief in the special relevance that evolution had for our potential to actively control the characteristics of members of future human generations. As we have seen, here Galton appealed not so much to natural selection as to the artificial selection practiced by animal and plant breeders. It was artificial selection that Darwin had introduced in his powerful, founding analogy for the mechanism of natural selection in the first chapter of *On the Origin of Species by Means of Natural Selection*. Natural selection was, for Darwin, like artificial selection *without a personified artificer*. Galton proposed that increasing scientific knowledge and novel technologies could inform an artificer or selecting agent sufficiently to allow control of the characteristics of future generations of people. In short, Galton held that science and technology would allow us to determine, both in theory and practice, what sorts of people there should be.

Galton long believed that the mental characteristics and abilities we possessed as a species were particularly important for our past, present, and future survival, and he sought to bring them into the Darwinian fold. Galton was impressed both by what he considered to be extreme variations within the human species on this dimension and by the intra-familial patterns that he saw within this variation. He was especially drawn to a kind of clustering of mental talents and abilities within certain families, including his own. These patterns were the basis for Galton's belief in the heritability of mental talent and ability, where this heritability was understood to have a biological basis.

Galton's eugenic thinking first saw the light of day in a pair of articles, "Hereditary Talent and Character" published in *MacMillan's Magazine* in 1865, which were then expanded into a book, *Hereditary Genius*. In both this early work and in his *English Men of Science: Their Nature and Nurture* Galton advanced ideas that can be readily seen as part of the short history of eugenics, despite introducing the term "eugenics" only in a later book, *Inquiries into Human Faculty and Its Development*. Searching for a word to replace "viriculture," which he himself had previously introduced and used, Galton says that we

greatly want a brief word to express the science of improving stock, which is by no means confined to questions of judicious mating, but which, especially in the case of man, takes cognizance of all influences that tend in however remote a degree to give to the more suitable races or strains of blood a better chance of prevailing speedily over the less suitable than they otherwise would have had.

"Eugenics" was that word, and Galton's first characterization of eugenics in terms of improvement, judicious mate choice, and the prevalence of the "more suitable" cohered with his earlier talk of talent and genius in positive terms.

In articulating the idea of eugenics, Galton used basic tabulations and statistical analyses to buttress the common observation that talents ran in families. Common sense and fairly elementary mathematical calculations coalesced to support the idea that there was a scientific case to be made for human improvement through the positive selection of traits. Over the next thirty years, however, "eugenics" was a word that not only continued the positive evaluative tone that Galton set in finding a place in the names of influential societies aiming at human or "race" betterment. That word was also woven into the fabric of discussions of the social problems of criminality and poverty (or "pauperism") and their solution. This eugenic attention to such social problems motivated specific pieces of legislation concerning marriage constraints, reproductive sterilization, and immigration restrictions in both North America and Europe. Eugenics was to be a science concerned with improving the "more suitable." Yet from the outset eugenics was also to be the basis for removing the "less suitable" from future generations.

One of Galton's own final statements about eugenics is blunt about these positive and negative forms that eugenics might take, giving priority to the latter. In the concluding paragraph to his reflective "Race Improvement," the last essay in his *Memories of My Life*, Galton says that eugenics' first object is to check the birth-rate of the Unfit, instead of allowing them to come into being, though doomed in large numbers to perish prematurely. The second object is the improvement of the race by furthering the productivity of the Fit by early marriages and healthful rearing of their children.

Thus, Galton's proposed science, eugenics, was to form one of the fragile sciences, one integratively drawing on putative facts from each of the biological, cognitive, and social sciences to promote the betterment of human society over generations. Yet it was also a fragile science whose ramifications for some parts of society were not so hope-inducing. The division between "the Fit" and "the Unfit" permeated eugenic thinking from the outset, and that division set a dual track—the positive and the negative—for social policies with the laudable-sounding aim of human improvement.

By reading past the quaintness occasioned by Galton's original talk of "improving stock" or "strains of blood," and appreciating that the science he had in mind was to be applied science of a special sort, we can identify three key components to this initial Galtonian conception of eugenics. These three components specify the *what*, the *why*, and the *how* of eugenics, jointly giving us a characterization useful for thinking about both the history and the contemporary status of eugenics. What is eugenics? In the Galtonian vision at least, eugenics is

(a) a systematic set of ideas, practices, and policies putatively grounded in scientific knowledge and technology that (b) aims to improve the quality of human lives across generations (c) by changing the designated composition of particular human populations to produce more desirable and/or fewer undesirable sorts of people.

I want to suggest in what follows that The Eugenic Mind is best understood in terms of this Galtonian vision.

Both in public and popular culture as well as in historically focused academic work, eugenics is often associated with two elements that are missing from this three-part characterization of eugenics: state control or regulation of individual reproductive choice, and coercive interference in the lives of individuals. Eugenic thinking, however, transcends moments and trends marked by those features, prominent and important as they were during the second half of the short history of eugenics. In the next section I briefly unpack and comment on the three components of the Galtonian vision of eugenics to further understand The Eugenic Mind and how eugenics was positioned between science and social movement.[2]

2.3 Eugenics as Applied Science

Identifying eugenics as applied science may be thought to imply very little, saying only that eugenics does not fall under the contrasting mythical category of "pure science." But the labeling of eugenics as applied science should be taken not so much to register a location on the putative divide between pure and applied science as to distinguish eugenics from a certain idealization of scientific inquiry. It signals three things that eugenics is *not*, and never was: it is not merely theoretical, not primarily mathematical or statistical, and not value-free.

First, eugenics is *not merely theoretical*, in the sense of being concerned primarily with abstract or idealized conditions (cf. theoretical physics or theoretical biology). It is focused on, and very much motivated by, perceived problems in real-world human populations and their solution.

Second, eugenics is *not primarily mathematical* or statistical in nature, however much it may at times draw from or rely on mathematical techniques or results. Galton himself was an accomplished mathematician, inventing several statistical techniques, such as the quantified idea of a standard deviation and the use of regression lines in statistics, which remain with us today. Galton's most prominent successors in the United Kingdom—Karl Pearson and Ronald Fisher—were also statistically sophisticated innovators who led a biometric wing to the eugenics movement. While the quantitative measurement of both individuals and populations has played an important role in the short history of eugenics, much eugenic work bears no closer a relationship to the underlying statistics than does the bulk of contemporary biological, cognitive, and social sciences.

Third, eugenics is *not value-free science*, and doesn't purport to be: it is deeply and often explicitly value-laden. I want to take a little more time to explain this dimension to the applied nature of eugenics, for doing so will take us to some core aspects of eugenics as a mixture of applied science and social movement.

Most obviously, eugenics rests on a judgment about what makes for more desirable or better human lives, and thus on views of what it is that constitutes an improvement to human life. To improve human life or to make human lives better, we need to know what makes a human life good, in some sense, and what makes a human life bad in that same sense. At least some such value judgments may be relatively uncontroversial or widely agreed to. For example, there is widespread (even if not complete) agreement that a human life full of severe pain and suffering, other things being equal, is worse than a human life that has at most sporadic and minor pain and suffering. This value-laden dimension to eugenics is no more problematic than the value-ladenness of medical practices that aim to cure diseases, restore previous functioning, or alleviate pain. The value-ladenness here is as apt as it is in the applied science of structural engineering when that science fosters projects that aim to build bridges that are stable, durable, and resilient in the face of local geological and weather conditions. Nobody rationally wants a bridge that will collapse.

But eugenics is value-laden in two other ways that go beyond this kind of relatively unproblematic form of value-ladenness. These dimensions to the value-ladenness of eugenics continue to fuel The Eugenic Mind. Indeed, the ongoing interaction between them makes for a heady, emotional cocktail in contemporary politics and public life, particularly in the United States.

First, the evaluative judgments that The Eugenic Mind rests on go well beyond those for traits, behaviors, and characteristics whose desirability or undesirability can be properly taken for granted. Second, eugenic thinking presumes that there are more desirable and more undesirable—better and worse—kinds or sorts of people. For this reason, the primary way in which eugenics has sought to improve the quality of human lives over generational time has been by advocating for ideas and policies that promote there being a greater proportion of *better kinds or sorts of people* in future generations.

To illustrate the first of these points, many eugenic policies were either explicitly stated in terms of, or implicitly relied on, a positive valuation of high intelligence and a negative valuation of low intelligence, especially as measured by standardized IQ tests, such as the Stanford-Binet. While this positive valuation of intelligence is still widely shared in our society when expressed abstractly, as part of a science that aims to inform and shape what sorts of people there should be in future generations, it is a value judgment that is significantly more questionable than that concerning the avoidance of pain and suffering. Eugenic thinking and practice also rested on assessments of a broader range of personality and dispositional tendencies—for example, clannishness, cheerfulness, laziness, honesty, and criminality—not only whose transmissibility across generations was considered controversial but whose very existence as intrinsic traits and tendencies has never had substantial scientific support.

Likewise, turning to the second point, the eugenic thinking that informed immigration policy in the United States following the First World War held that people of different races or ethnicities were differentially desirable as immigrants coming into the country. This differential valuation was applied to groups such as Poles, Greeks, Italians, Jews, and Slavs. Thus, eugenic immigration policies aimed to promote the influx of immigrants who were viewed as more desirable in nature, and to restrict the immigration of those deemed to be of inferior stock. We now question whether such groups of people are properly thought of as more or less desirable sorts of people to produce future generations of American nationals. But we also rightly wonder whether these are *sorts* of people, in the relevant sense, at all.

Apart from reminding us of the contemporary resonance of The Eugenic Mind, these examples indicate the way in which the aim of eugenics as applied science—to inform and direct human improvement over generational time—was not only meliorative but was to be achieved in a particular manner. After all, one way to so improve the quality of human populations over generations would be to promote ideas, policies, and practices that aimed to ensure the realization of *every* person's potential to produce and raise children with at least as much "quality" as their parents. Many public health measures—such as reducing the amount of lead in home and school environments that would otherwise likely negatively affect children's developmental processes—are of this nature.

That was not, however, the way of human betterment favored by the applied science of eugenics and that continues to forms a key part of The Eugenic Mind. Instead, historically eugenicists typically followed Galton in emphasizing that quality was not equally distributed in the kinds of human populations that are regulated by governmental policies and jurisdictional legislation. More specifically, they thought of such populations as being composed of fundamentally distinct kinds of people, with some kinds being of higher quality than others. Some of these sorts of people were to be improved through eugenic policies that encouraged their reproduction; others were to be eliminated over generational time. The goal of intergenerational human improvement within the eugenics movement was thus to be achieved by increasing the proportion of higher-quality people in future generations, and this could be achieved in two ways under eugenic logic. Thus, eugenicists historically advocated ideas, laws, policies, and practices either that aimed to maximize the reproduction of higher-quality people—positive eugenics—or that aimed to minimalize the reproduction of lower-quality people—negative eugenics. Or both.

This dimension to The Eugenic Mind remains very much with us. The ugly us-versus-them racial and ethnic divisiveness that depicts an inferior Eugenic Other as the deep root of contemporary social problems is just its most obvious manifestation. As we will see in the remaining chapters in part I and in part II, The Eugenic Mind shows those same dimensions of value-ladenness in contemporary engagements with disability and human variation in biomedical and biotechnological contexts. The distinction between the more favorable and the less favorable "strains of blood" that we find in Galton's original definition of eugenics—a divide that was often cast in terms of the bifurcation between "the Fit" and "the Unfit" in eugenics in the first half of the twentieth century—is not simply a legacy of a eugenic past. It is a hallmark of The Eugenic Mind itself.[3]

2.4   Between Science and Social Movement

Recognizing these dimensions to the value-ladenness of eugenics as an applied science among the fragile sciences should give pause to the naïve thought that eugenics was envisioned *simply* as science. To characterize eugenics only as a science would express its own kind of naïveté. For eugenics was also, particularly in the first half of the twentieth century, a fervent social movement, one championed by many community leaders—from politicians to university presidents and provosts, from farmers to members of elite and learned societies, from radical progressives to political conservatives.

The most manifest early influence of eugenics was temporally and spatially distant from its Galtonian origins in Great Britain in the 1860s. Instead, in the early twentieth century, eugenics gained a foothold both across the English Channel in continental Europe and across the Atlantic in the United States. By the end of its first decade, academic work on "racial hygiene" by Alfred Ploetz and Friedrich Hertz in Germany led to societies and institutions in France, Denmark, and Germany that advocated eugenic practices. By that same time, a handful of American states had followed the lead of the state of Indiana in 1907 in passing compulsory sterilization legislation applying to the "feeble-minded." Thus, eugenics also needs to be understood historically as a social movement, one that appealed to and recruited much of the leading science of its day to advance its aims and aspirations. The short history of eugenics is, in part, a short history of this social movement.

Around the time that Galton was attempting, as he put it, "to prove" that talent and ability were heritable by examining the patterns of reputation, accomplishment, and profession in elite British families, concerns about expanding reliance on public charity in New York State gave rise to the first of what became an influential series of family studies. This first study focused on the family given the name "The Jukes." Based on public health work in the mid-1870s by Elisha Harris, a doctor specializing in infectious diseases who later became the corresponding secretary of the New York Prison Association, Richard Dugdale's *The Jukes: A Study in Crime, Pauperism, Disease, and Heredity* was motivated by a concern with understanding poverty, criminality, and alcoholism. The study itself focused on constructing genealogies of forty-two families that could be traced to a common ancestor. A supposedly disproportionate number of these family members found themselves in prison, engaged in behavior deemed criminal or immoral, and were economically impoverished.

Despite Dugdale's own resistance to hereditarian explanations of the genealogies that he identified, *The Jukes* became a model for future studies— The Tribe of Ishmael, The Kallikaks, The Nam Family. Collectively these studies, eleven of which the criminologist Nicole Rafter later reprinted together under the title *White Trash: The Eugenic Family Studies*, served to reinforce the view that social problems of prostitution, theft, murder, and poverty had a hereditary basis. Such studies, many of which were funded directly by the Eugenics Record Office at Cold Spring Harbor and published between 1912 and 1926 by members of its staff, most notably Arthur Estabrook, and by the superintendent of Vineland Training School for Feeble-minded Girls and Boys, Henry Goddard, played an influential role both in the science of eugenics and in the social movement of eugenics that this research fed (see chapter 3).

As implied in chapter 1, the eugenics movement has been well studied, particularly by historians. The rise of eugenics in the United States has been especially thoroughly discussed, as have the eugenics movements in Great Britain, Germany, and Scandinavia. Work here has included studies of leading eugenic organizations, eugenics in specific national contexts, and the ways in which particular sciences and technologies—for example, genetics, psychology, sterilization—functioned in eugenic thinking and practice. In recent years, historians and other scholars have looked beyond these now familiar contexts for eugenics and aimed to contribute to a world history of eugenics that offers a more comprehensive view of the short history of eugenics. Alison Bashford and Philippa Levine's impressive edited volume, *The Oxford Handbook of the History of Eugenics*, is representative of this ongoing trend in work on eugenics, structured both to provide a more geographically inclusive range of national contexts for eugenic movements and to trace transnational themes across the various eugenics movements.

This expansive approach to understanding eugenics as a social movement contains mutually directive "in" and "and" dimensions. Eugenics is to be understood *in* specific national and regional contexts: eugenics in Australia, Canada, the United States, Eastern Europe, the Netherlands, South Africa, for example. Eugenics is also to be explored in terms of its relationship to something else, whether that be a discipline, a view or idea, a historical trend, or a social phenomenon: eugenics *and* genetics, sexuality, race, fertility control, the Darwinian context, colonialism, psychiatry, internationalism, genocide—to sample from the relevant chapters from Bashford and Levine's *Handbook*.

Both dimensions have made our understanding of eugenics as a social movement more sophisticated, allowing us to probe further into The Eugenic Mind, past and present. While the first dimension invites comparative questions and analyses about variations among and trends across different national contexts—for example, why did Canada enact eugenic sterilization laws while Great Britain did not?—the second dimension readily enriches our picture of the place of eugenics in the broader landscape of ideas, trends, and movements—for example, how is eugenics related to race and racism?[4]

2.5   Eugenics, Race, and Ethnocentrism

In the highly racialized world of the late nineteenth century, eugenic ideas were often formulated with respect to the kinds of people picked out by racial or ethnic groups. Yet this was not simply a racialized world. It was one with a long past of racism and ethnocentrism, both fed by a history of colonialism. It was a world in which the developed Western nations in which eugenics originated and took root had, over generations, systematically extracted and exploited resources, including people, from other parts of the world. Thus, for eugenicists, there were not simply various different racial or ethnic groups, but higher quality (superior) and lower quality (inferior) groups identified as races, tribes, or ethnicities.

As we have already seen, this eugenic thinking directed social policies concerning the geographical boundaries of national populations, such as immigration policies. For example, in response to the popular view that the United States was being overrun with undesirable immigrants, the 1924 Immigration Restriction Act (also known as the Johnson-Reid Act) reduced the existing quota on the percentage of immigrants to the United States coming from countries in southern and eastern Europe. At least part of the thinking underlying not only the act but the kind of existing quota-based system that it tightened was eugenic in nature. That thinking rested on the distinction between desirable and undesirable stocks of people in articulating a policy governing future generations, reinforcing the fear of a kind of degeneracy in the American population.

Indeed, prominent figures in the eugenics movement in the United States, such as Harry Laughlin and Lothrop Stoddard, played active roles in shaping the character and content of the Act, including as "expert eugenical witnesses" to the House Committee responsible for the Immigration Restriction Act. As director of the Eugenics Record Office at Cold Spring Harbor, Laughlin conducted surveys of state institutions for the feeble-minded between 1914 and 1919, testifying that Southern and Eastern Europeans were disproportionately represented in such institutions. Stoddard's popular 1920 book, *The Rising Tide of Color Against White World-Supremacy*, explicitly argued for immigration restriction laws based on a view of racial superiority and inferiority, and tied national degeneracy to "mongrelization," or the mixing of superior Anglo-Saxon and Nordic races with the "yellow" and "brown" races. The Immigration Restriction Act itself was rescinded in the United States only in 1965.

While calls for immigration restriction need not be eugenic in nature, those that share in the rhetorical divisiveness of eugenics past should at least trigger one's eugenics detectors. Yet the articulation of eugenic science in terms of two dimensions to the value-ladenness of The Eugenic Mind—the appeal to questionably transmissible and dubiously intrinsic traits and characteristics, and the shift from traits of people to kinds or sorts of people—provides a basis for more than alertness here. It allows one to assess the extent to which contemporary talk not simply of immigration but also of *the defective quality of immigrants* crosses a eugenic threshold. That talk should, correspondingly, also give us pause for more than chin-scratching reflection.

In many countries with a colonial past, eugenics also influenced government policies focused on nonimmigrant ethnic and cultural groups, such as aboriginal and native peoples. For example, "racial hygiene" organizations were formed in each of Denmark, Norway, and Sweden in the first decade of the twentieth century, and a central concern of each was the character of the existing populations in each country. In Sweden much of the work of the State Institute for Racial Biology, founded in 1922 and directed by the physician Herman Lundborg, was devoted to describing the racial character of the Swedish people. Here there was a specific focus on people in the north, especially the indigenous Sami people, who were deemed inferior and seen as posing a polluting threat to the Swedish nation through miscegenation.

In Australia and Canada, indigenous peoples were subject to "civilizing missions" that involved forced assimilation through the removal of children from their parental homes—to residential schools in Canada, and to government settlements and missions in Australia. These assimilationist policies were forms of eugenics through nonbiological means, forms of what we might call *cultural eugenics* that sought to rid the population of less desirable cultural variants, those expressed in indigenous ways of life.

In Canada, the residential school system formally resulted from the amalgamation of the systems of industrial and boarding schools in 1923, with residential schooling being made compulsory for children of aboriginal descent from 1920. In Australia, the absorption of light-skinned aboriginal children into white culture became part of explicit government policy in the 1930s, with the removal of indigenous children from their families from the age of four becoming state policy enforced via the existing protectorate system through the agency of the Chief Protector or Protection Board in each state and territory from 1940. In both countries, these early twentieth-century policies were based on the view that indigenous peoples were inferior, and that this kind of intergenerational intervention would produce, over generational time, a solution to a perceived social problem. What in Canada was called the "Indian problem" was to be solved by a form of culturally mediated eugenic intervention; in Australia, the legacy of these policies has been discussed in terms of the effects of and on the "stolen generations." Thus, the relationship between eugenic thinking and racism and ethnocentrism is both manifest in national immigration policies and implicit in assimilationist policies directed in part at indigenous peoples within national boundaries.

However, race and ethnicity were often not the most salient categories in terms of which eugenic ideas, policy, and legislation were articulated. Despite the centrality of such categories to the colonial dimensions of eugenics and immigration policies, this has created some ambivalence—perhaps even confusion—about the relationship between eugenics and race and racism. For example, consider what Bashford and Levine say about this relationship in their introduction to the *Oxford Handbook of the History of Eugenics*. They say that eugenics and racism

> have become almost interchangeable terms, but the association is perhaps too simplistic. ... [M]uch, if not most, eugenic intervention was directed at "degenerates" who already "belonged," racially or ethnically, "internal threats" or "the enemy within" whose continued presence diluted the race. ... To be sure, these were projects of racial nationalism and indeed racial purity—eugenics was never not about race—but the objects of intervention ... were ... marginalized insiders whose very existence threatened national and class ideals.

When we turn to those "marginalized insiders," while race and ethnicity are sometimes relevant variables, they are not the primary traits through which the social mechanics of eugenics operated within provincial and state-level jurisdictions, as we will see more clearly in chapter 3. In many such jurisdictions, the sorts of people who were subject most directly to the eugenic gaze and subject to eugenic intervention for the supposed menace they posed to society were the *mentally deficient*, referred to as *mental defectives* and *the feeble-minded* throughout most of the short history of eugenics. In fact, the mentally deficient or feeble-minded quickly became the most significant sort of people targeted by eugenic laws and policies. They remained so throughout the short history of eugenics.[5]

2.6   Galton, Mental Abilities, and the Weak-Minded

As eugenics established itself by the end of the first decade of the twentieth century both as a putative applied science and a full-blown social movement, of the various sorts of people lumped together as unfit for present or future society, those thought to have a mental deficiency became an increasing focus. Indeed, by that time, the weak- or feeble-minded were viewed as playing a key role in generating the social problems to which eugenics was, in part, a response. As the physician and specialist on mental deficiency, Alfred Tredgold, said in editorializing on the recommendations of the British Royal Commission on the Care and Control of the Feeble-Minded in 1910, the feeble-minded are

kith and kin of the epileptic, the insane and mentally unstable, the criminal, the chronic pauper and the unemployable classes, and I am convinced that the great majority of the dependent classes existing today owe their lack of moral, mental and physical fibre to the fact that they are blood relations of the feeble-minded and are tainted with their degeneracy.

This eugenic focus on the feeble-minded was the foundation for a range of ideas and policies that fall under what the sociologist Nikolas Rose has called *psycho-eugenics*, whereby psychological characteristics became both diagnostic of degeneracy or unfitness and the key to the pursuit of social strategies of intervention. This focus was seeded in Galton's own early eugenic thought. As Rose argues, in addition it not only permeated the views of physicians and others within the medical establishment, but also shaped the trajectory of the discipline of psychology.

Galton's call for the integration of *Homo sapiens* into a Darwinian view of the living world, as we have seen, formed part of his motivation for advocating eugenics. Our own species is subject to the laws and regularities governing intergenerational stability and change more generally in the plant and animal world. As we have also seen, Galton was especially interested in ensuring that this integration recognize an individual's *mental* characteristics, and not just his or her physical features, as subject to the forces of selection, whether natural or artificial.

Yet this was not primarily a matter of parity, a matter of ensuring that mental traits be treated alongside physical traits. For Galton also clearly recognized the mental aspects as especially relevant in determining an individual's nature and what that individual could contribute to future generations. As even the titles of both *Hereditary Genius* and *Inquiries into Human Faculty* make clear, when Galton considers the superior abilities of some people and their relevance to intergenerational improvement, he is focused on mental and intellectual abilities, "genius" and "faculty" being associated, then as now, with the human mind. In his original preface to *Hereditary Genius*, Galton notes that the topic of the book arose "during the course of a purely ethnological inquiry into the mental peculiarities of difference races." In the preface to the book's second edition, published after Galton had coined "eugenics" as a term, his acknowledgment that "ability" would have been a better choice of term than "genius" in his original title serves to clarify that Galton sees "natural ability and intelligence," to use a phrase that recurs in his writings, as warranting eugenic attention.

The principal aim that Galton had in these works was twofold. First, to show that such mental abilities were heritable; and, second, to convince others that the natural variation of mental abilities in a population was subject both to degenerative tendencies—if left to natural selection—and to meliorative tendencies, if instead subject to certain forms of artificial selection.

Galton's pioneering work was clearly centered on the latter of these, advocating practices that would later be known as *positive* eugenics for those taken to possess superior abilities. Those with diminished abilities, however, are recognized in the broader social vision that Galton conveys in these early works. In concluding the penultimate chapter to *Hereditary Genius*, Galton sketches this vision:

The best form of civilization in respect to the improvement of the race, would be one in which society was not costly; where incomes were chiefly derived from professional sources, and not much through inheritance; where every lad had a chance of showing his abilities, and, if highly gifted, was enabled to achieve a first-class education and entrance into professional life, by the liberal help of the exhibitions and scholarships which he had gained in his early youth[;] where the weak could find a welcome and a refuge in celibate monasteries or sisterhoods, and lastly, where the better sort of emigrants and refugees from other lands were invited and welcomed, and their descendants naturalized.

Here Galton's "welcome and refuge" for the weak, especially the weak-minded, is complemented by a sense of moral deservingness for at least some such weak-minded people, namely those who recognize the potentially degenerative effects of their own reproduction. In his "Conclusion" to *Inquiries into Human Faculty*, Galton continues with his contrast between the highly gifted or better adapted, on the one hand, and the weak or "the rest," on the other:

> The stream of charity is not unlimited, and it is requisite for the speedier evolution of a more perfect humanity that it should be so distributed as to favour the best-adapted races. I have not spoken of the repression of the rest, believing that it would ensure indirectly as a matter of course; but I may add that few would deserve better of their country than those who determine to live celibate lives, through a reasonable conviction that their issue would probably be less fitted than the generality to play their part as citizens.

The praise that Galton gives to that small portion of "the rest" in a position to willingly choose celibacy may catch our attention here. But it should not distract us unduly from what Galton notes he has not previously spoken of: the "repression of the rest."

Galton's own texts reveal little about the way that those more commonly referred to as "idiots" and "the feeble-minded" were viewed in Great Britain at the time. We need to look elsewhere to understand the social context in which views of mental deficiency shaped, eugenics. From all of Galton's talk of individuals, ability, and celibacy, it would be easy to forget or overlook that *children*, especially "defective children," became a particular target of eugenicists as their ideas became policy and practice.[6]

2.7   Eugenics and the Mentally Deficient

Victorian England at the time that the short history of eugenics began in the 1860s was undergoing an expansion in governmental facilities for lunatics. These institutions housed diverse groups of people: children and adults; violent criminals and the physically impaired; those infected with tuberculosis and those with sensory losses, such as blind and deaf people; those thought to have hereditary mental deficiencies and those whose mental limitations were known to be the result of life events, such as workplace accidents and fights. The final decades of the nineteenth century saw two important changes as part of this expanding state role in the lives of people institutionalized at lunatic asylums.

The first was an increasingly fine-grained set of classifications of people thought of originally under the heading *lunatics*. This included the distinction between affective, moral, and ideational insanity in the early work of Henry Maudsley, and the distinction between kinds and grades of what had until then generically been referred to as "idiocy." For example, among those deemed mentally deficient or feeble-minded there were now *low-grade* idiots and *mid-grade* imbeciles—later extended to include the *high-grade* feeble-minded who would get their own specially coined term in the early twentieth century across the Atlantic in the work of Henry H. Goddard: morons. The foundational distinction for this proliferation of sorts of people suffering from mental diseases, conditions, or problems was that between *idiots*, who possessed generally low levels of mental functioning, and *lunatics*, whose behavior was regularly or even just occasionally governed by abnormal mental states, such as hallucinations, delusions, extreme and negative emotions, and impactful losses of self-control.

This evolving distinction—in effect between what would become *the feeble-minded* and *the insane*, between those who were mentally deficient or defective, on the one hand, and those who suffered from mental illness or disorder, on the other—played an important role in The Eugenic Mind and in eugenic policies and legislation. For while both idiots and lunatics were thought to pose threats to present and future society, those threats were of very different kinds and their patterned manifestations distinctive from one another. Consider four such contrasts between insanity and idiocy.

Insanity manifested itself in readily observable self- or other-destructive behaviors, arose during late adolescence or adulthood, was seen frequently enough to be curable or treatable in principle (if not always in practice), and was often manifest in a person's life only in particular contexts or during specific periods. *Idiocy*, by contrast, was often more subtly manifested and thus often required special training or technologies to be detected, was present at birth or early in life while the individual was still undergoing development, did not readily admit of a cure or rehabilitation, and pervaded an individual's life and so was viewed as imposing a severe constraint on what that individual could accomplish in any life endeavor.

The threat that lunacy posed was thus primarily immediate and present-directed, but the individuals posing the threat were themselves remediable; the threat posed by idiocy, by contrast, was primarily longer-term and future-directed. And idiots suffered from irremediable defects, from subnormalcies that could not be made normal, from an inborn or developmental derailment from normalcy that could not be set right. At least that is how the distinction between insanity and feeble-mindedness was first drawn, even if the contrast here softened over time. As hereditarian views of the mind strengthened within the eugenics movement, insanity came to be seen as irremediable in much the way that feeble-mindedness was. And conversely, as panic developed about the putatively expanding prevalence of both types of mental unfitness under welfarist social policies, the threat posed by the feeble-minded came to be viewed as immediate as that posed initially by the insane.

These developments in the conceptualization of the mentally dysgenic were accompanied by a second change, one in the practice of institutionalization. This was the recognition of the need to construct, partition, or structure institutions in accord with the particular subpopulations of this unfit morass of humanity. Putatively feeble-minded children came to be perhaps the most significant of these groups within institutions. This was in part because of the rise of compulsory public education and the question of the most appropriate form of education or "training" for the feeble-minded child. This question was addressed in publications by those working at training schools, such as Martin Barr, the long-serving chief physician at the Pennsylvania Training School for Feeble-Minded Children. Since the resultant, costly expansion of institutional facilities that effecting this change would require was not matched by correspondingly greater fiscal outlays for such institutional changes, this placed pressures on existing institutions and those responsible for them: from superintendents and other institutional staff through to politicians and community leaders. It is in this context that feeble-mindedness was constructed as a *problem*.

As the historian James Trent has shown, in the United States during the second half of the nineteenth century there was a transformation in how this problem of feeble-mindedness was conceptualized. After the 1840 U.S. census, in which "idiots" had been counted in North America for the first time, idiocy became more widely recognized as linked to a range of social problems, including poverty and crime. Between the time of the census and the end of the American Civil War in 1865, mental deficiency and feeble-mindedness came to be viewed as root causes of such problems requiring state-level resources to resolve. Here segregating the feeble-minded through their institutionalization in dedicated "training schools" came to occupy a more central role as a response to that problem. After the American Civil War, as eugenic ideas about intergenerational social melioration were articulated about undesirables—criminals, paupers, idiots, the insane during the last quarter of the nineteenth century—feeble-mindedness began to be viewed as the core, underlying heritable defect whose removal from the population would reap broader social benefits. But it is only with the rise of eugenics as a social movement in the United States in the early twentieth century that feeble-mindedness came to be viewed as an inherent *menace to society*. This menace was neither perceived as stemming from social conditions, nor to be countered through educational improvement. Rather, feeble-mindedness itself became biomedicalized, and the fragile science of eugenics was the key to its study and elimination.

Around the time that Arthur Tredgold's work at the administrative interface between medical expertise in mental health and government policy in Great Britain was singling out feeble-mindedness as the root cause of social dependence, across the Atlantic in North America eugenic thinking began to occupy another administrative interface that consolidated that same identification. The Harvard-trained zoologist, Charles Davenport, directing the newly formed (1912) Eugenics Record Office at Cold Spring Harbor in the United States, was looking to address the social concerns that had been articulated by the family studies stretching back to Dugdale's *The Jukes* thirty-five years earlier. Despite lacking expertise in "mental defect," Davenport readily appealed to the growing idea that feeble-mindedness was a kind of key to human melioration, reporting that

it is now coming to be recognized that mental defect is at the bottom of most of our social problems. Extreme alcoholism is usually a consequence of a mental make-up in which self-control of the appetite for liquor is lacking. Pauperism is a consequence of mental defects that make the pauper incapable of holding his own in the world's competition. Sex immorality in either sex is commonly due to a certain inability to appreciate consequences, to visualize the inevitableness of cause and effect, combined sometimes with a sex-hyperaesthesia and lack of self-control. Criminality in its worst forms is similarly due to a lack of appreciation of or receptivity to moral idea.

The appeals here to evils of the adult world—alcoholism, pauperism, sexual immorality, and criminality—mask a point about the targets of eugenics that I made at the end of section 2.6. At the heart of the menace to society posed by the eugenic threat of feeble-mindedness were putatively mentally defective *children*. What to do with those defective children who already existed, and how to avoid producing more of their kind in the future, remained frontline problems to which eugenic thinking was proposed as a solution. As we move on in chapters 3 and 4 to further explore the social mechanics of eugenics, we will return to this aspect of The Eugenic Mind.[7]

2.8   The Long Past of Eugenics

I began this chapter with Ebbinghaus's well-known contrast between the short history and long past of the discipline of psychology, and then focused the core of the chapter on eugenics' short history. I have suggested a view of The Eugenic Mind both as having been forged among a plurality of fragile sciences in the second half of the nineteenth century. Eugenics acquired the mixed status of applied science and social movement in part as a function of that location and in part due to its central aim of human improvement. I have argued that those deemed to suffer mental deficiency and abnormality were, from the outset, the principal targets of The Eugenic Mind, a view that I will look to further support in the next chapter in turning to discuss the idea of a eugenic trait.

Innocuous-sounding talk of human improvement, betterment, or melioration—particularly with respect to the mental and cultural endowments we leave to future generations—does not immediately call to mind the harsher realities of eugenic policies and practices. The dissonance between the public face of eugenics past and those harsher realities itself should serve as a caution in reflecting on contemporary Eugenic Mind Projects, present and future. But let me conclude here by shifting from that short history and its significance to eugenics' long past.

Talk of what I am calling the long past of eugenics is commonly found in the works of those who sought to justify eugenic ideas and practices during eugenics' short history. For example, invoking the long past of eugenics has this role in Edgar Schuster's *Eugenics* and in Allen G. Roper's *Ancient Eugenics*, both published shortly before the First World War. Prominent in such works are appeals to ancient Greek philosophers, especially Plato, particularly *The Republic* and *The Laws*. While nobody quite adapts White-head's famous quip to suggest that Galton's eugenics is simply footnotes to Plato, locating modern eugenicists against the background of this longer, veritable eugenic past serves to convey the idea of a tradition of intellectual exploration that has been revived as an applied science of contemporary value and significance. In this respect, it functions much as does Ebbing-haus's appeal to the long past of psychology.

Like Ebbinghaus's influential appeal—which in fact mentions only Aristotle and Heraclitus in his intimations of the ancient basis of psychol-ogy as a field of inquiry—these appeals to the long past of eugenics in fact have little substance to them, whatever their rhetorical role. Consider the examples of Schuster and Roper. Roper points to practices of infanti-cide in Sparta and occasional quotations from classical authors drawing analogies between better breeding in animals and human reproduction. Schuster provides an extensive discussion of Plato's views on what we might think of as eugenically related themes, such as preferential sexual partnering for reproduction and the disposal of less worthy infants in his ideal republic.

Although in this chapter I have had little to say that is standpoint-ish, I want to finish here by returning to how this same kind of appeal to eugen-ics' long past also found expression in Alberta's eugenics history. For Provost John MacEachran also appealed to Plato to provide backing for the eugenic dimension to the mental hygiene movement in his "A Philosopher Looks at Mental Hygiene." MacEachran frames his discussion in terms of the idea of individual and collective *catharsis*, a kind of purification or cleansing or the individual soul or the collective way of living. MacEachran draws on Plato directly in two ways, initially lightly and passingly, by quoting Plato's *The Republic* approvingly on the social futility of setting out to "lengthen good-for-nothing lives" and the undesirability of "weak fathers begetting weaker sons." MacEachran then goes on to echo Schuster's discussion in turning more explicitly to Plato's infanticidal view that "the offspring of the infe-rior, or of the better when they chance to be deformed, will be put away in some mysterious, unknown place, as they should be" and the general regulation of marriage and parenthood by wise, ruling philosopher kings. MacEachran draws the following conclusion:

We may not, perhaps, be prepared to go as far as Plato recommends in the way of restricting marriage and the procreation of children; but it is well to recognize that about twenty-five hundred years ago the greatest thinker in the western world was giving the most careful consideration to problems that we, in spite of our much-vaunted progress and efficiency, have scarcely attacked or even seriously ventured to discuss in public.

It seems likely that MacEachran was in part inspired and influenced by a noble vision of eugenics, one that he sees rooted in "the greatest thinker in the western world," in his own role as chair of the Alberta Eugenics Board, a position that he had accepted just a few years previously. Perhaps, as the psychologist Doug Wahlsten has intimated, MacEachran even saw himself in the role of a minor philosopher king regulating at least the production of children. But like other gestures toward the long past of eugenics, missing from MacEachran's brief foray into that putatively longer past is any substantive discussion of what eugenic knowledge and technologies would encompass, and little more than an appeal to a distant authority. If eugenics really does have a long past worth knowing about, then that past still remains to be uncovered.

As our discussion of eugenics in this chapter indicates, eugenics is the coalescence of a set of ideas—about sorts of people, human nature and variation, mental deficiency, moral and mental degeneration, fixed human types, cost effectiveness, and heritable transmission—and proposals for their realization articulated in the late-nineteenth-century fragile sciences. Those ideas percolated out in a mixed applied science/social movement whose existence delineates the short history of eugenics. That short history is one thing that marks off eugenics from a variety of "isms"—racism, ethnocentrism, ableism, sexism, for example—and allows us to understand the shifting and at times complicated relationships between these and The Eugenic Mind. The blending of applied science and social movement to systematically embed eugenic ideas into social practices and policies is, however, more than a systematic extension of preexisting eugenic thinking: it is, rather, what makes the short history of eugenics the only real past that eugenics has had.

More important, however, than a degenerating debate over the long past of eugenics, is the question of whether eugenics *is* past. From the standpoint of the putatively mentally defective, and perhaps of others who have been past targets of The Eugenic Mind, there is a decisive answer to this question. Before returning to this question in chapter 4, however, I want to probe further into the characterization of eugenics and The Eugenic Mind given in this chapter and develop a little more complexity to the view of the relationship between the theory and practice of eugenics. I do so by discussing a basic question so far largely unasked: What is a eugenic trait, exactly?[8]

3 Specifying Eugenic Traits

3.1 What Is a Eugenic Trait?

Throughout most of the twentieth century, some people were subject to eugenic classification and treatment based on their supposedly having certain traits. Those traits, such as intelligence and mental deficiency, were the focus of research, publications, and propaganda generated by pro-eugenic individuals and organizations, and found their way into marriage, immigration, and sterilization laws in North America and elsewhere. In the most extreme case—that of Nazi Germany—having one or more of those traits became literally a matter of life and death.

The most fundamental question to ask here is this: what is a eugenic trait? Answering this question should inform our understanding of the social mechanics of eugenics. Although there is a sense in which this question could arise naturally for anyone reflecting on eugenics, for me it was prompted by dissonance I felt when reflecting on an initially puzzling and troubling discrepancy. This was the discrepancy between eugenics on paper and the lived experience of survivors in Alberta, a gap between putative deficiencies of who was supposed to be targeted by past eugenic laws, policies, and practices, and the character of who in fact was so targeted. The striking nature of that gap only became apparent gradually over time, however, as the human edge to that discrepancy was made vivid to me by the life stories of eugenics survivors—the gossipy narratives of their ongoing day-to-day lives as much as their more reflective recollections of their institutionalized past. People like Leilani Muir, Judy Lytton, and Glenn Sinclair had been institutionalized, sexually segregated, and sterilized in Alberta because there was supposed to be something significantly wrong with them, something whose possible transmission to future generations was deemed serious enough to warrant bodily intervention in the form of sexual sterilization. Yet the engaged individuality of each of them, warts and all, sat well neither with any of them being the menace to society that they had been thought to be, nor with their classification as mental defectives and morons deemed unfit to parent. What were the eugenic traits that they possessed, or were supposed to possess, that justified this labeling, classification, and treatment? And what had gone wrong in the process that begins with ideas about human improvement and ends with a variety of individual human tragedies, including wrongful confinement and sterilization?

I was familiar with the phenomenon of wrongful accusation and conviction from several other contexts in which I had trafficked in my work life—most notably in cases of "satanic" (as it was originally called) or ritual (as it later became known as) child sexual abuse. Those cases, in which hundreds and sometimes thousands of bizarre accusations of large-scale child sexual abuse were levelled at individuals and groups of individuals, had become widespread in North America in the 1980s, leading not only to criminal charges but also to criminal convictions and long jail terms for people who had, at the end of the day, committed no relevant crime at all. Here was an instance where understanding what had really happened in such cases seemed to require taking an extremely marginalized standpoint—that of those accused as child molesters. For many, this was something as tough to do epistemically as it was morally, both during the unfolding of the many cases here and even long after.

So the abstract-sounding question "What is a eugenic trait?" was initially for me a question of how the traits identified as being of eugenic interest in the abstract came to wrongly identify people I knew increasingly as friends and members of my local community. That identification began a process that had devastatingly shaped the trajectories of their lives. Like standpoint theory more generally, standpoint eugenics has interwoven epistemic, metaphysical, and normative threads, each reinforcing the other. And again like standpoint theory more generally, the perspective afforded by standpoint eugenics generates novel lines of inquiry. Pursuing the idea of eugenics as a form of wrongful accusation was one of these, something I shall return to discuss in detail in chapter 8. Articulating the idea of a eugenic trait, and these simple questions about eugenic traits, represented another such line of inquiry for me.

As I have implied, the question "What is a eugenic trait?" was motivated by a certain kind of experience: local, particular, idiosyncratic. But we can also articulate this question independently of any particular standpoint and in a way that invites exploration further into the fragile science of eugenics itself. Many of the traits appealed to by eugenicists made their way into eugenic thinking from untutored, folk knowledge of characteristics of people. For example, "idiot" was a common term by the early nineteenth century, a term increasingly incorporated into the eugenic disciplining of the feeble-minded in the names of associations, organizations, and institutions. Characteristics or traits such as idiocy nonrandomly vary among people. Early in the short history of eugenics, it seemed to many that the systematic nature of this variation could be understood by appealing to a simple fact: that those traits ran in families. Given that some of those traits—like intelligence—were seen to be desirable, while others—such as mental deficiency—were seen as undesirable, eugenic laws and policies articulated by this science of human improvement would inevitably be focused on eugenic traits. Some of these traits were refined by eugenic research over time as new categories were formed and technologies for their more precise detection invented, underwriting policies of eugenic segregation, restriction, and sterilization. Many eugenicists believed these traits to be heritable in a stricter biological sense than their simply "running in families" and took the emerging science of genetics to facilitate the identification of eugenic traits. Which traits were these, more precisely?

Three chief sources reveal which traits were considered important to eugenicists, the public, the government, and individual decision makers: the research publications of eugenicists, discussions of eugenics in popular culture, and legislation introduced on eugenic grounds. While I shall focus on the last of these, particularly eugenic sterilization legislation, for reasons that will become apparent, I want first to say something briefly about the first two sources, research publications and popular culture.[1]

3.2   Research Publications

The research publications of proponents of eugenics include Francis Galton's early studies of genius, the "white trash" family studies that began with Richard Dugdale's *The Jukes* in 1877—both mentioned briefly in chapter 2—and the systematic research undertaken by the Eugenics Record Office (ERO) at Cold Spring Harbor under the direction of Charles Davenport from 1910. The investigation of family histories was central to all three, despite the different traits that each focused on.

As we saw in chapter 2, Galton's initial focus was on family lineages of positive achievement—on *hereditary genius* as he called it—showing that judges, statesmen, musicians, and wrestlers were found statistically clustered in a relatively small number of British families. Galton inferred that this clustering was due to those people having natural abilities that underpinned individual achievement status, with those natural abilities being transmitted hereditarily across generations. Even though Galton recognized from the outset that eugenics would need to attend differentially to both his "more suitable" and "less suitable" sorts of people, his evidential base for eugenics was built by making a case for the heritability of these more suitable "strains of blood" and the enrichment of human cultural achievement that would follow from positive eugenic selection. By contrast, the American eugenic family studies that *The Jukes* initiated focused more directly on those less suitable strains of blood. Such studies traced what would become known as intergenerational *social inadequacy,* focusing on poverty, criminality, and other social problems, and applying family pedigree construction to these traits. They were problem-centered and their aim was eliminative, both of the motivating social problems and of those who were seen to be their root cause.

The ERO sought to establish a more robust picture of eugenic traits through both research and public advocacy. Building on representations of family genealogies that had been nascent in early family studies such as *The Jukes*, the ERO invested its early efforts into developing a more systematic methodology that invoked standardized "charts of pedigree" that structured a series of publications. This series began with *The Study of Human Heredity* and *The Trait Book*, with Charles Davenport as the lead or sole author on most of these works. In the first of these, issued as Eugenics Record Office Bulletin No. 2, we can see how the ERO developed a standardized, simple representation of eugenic traits, built around an alphabetical ordering (figure 3.1).

From here, the systematization of an intergenerational map of eugenic traits progressed through the construction of a pedigree chart that sometimes used more elaborate lettering systems (e.g., double, lower-case letters

A	alcoholic, decidedly intemperate,	M	migrainous,
B	blind,	N	normal,
C	criminalistic,	Ne	neurotic,
D	deaf,	P	paralytic,
E	epileptic,	S	syphilitic,
F	feeble-minded,	Sx	sexually immoral,
G	gonorrheal,	T	tubercular,
I	insane,	W	vagrant (tramp, confirmed runaway).

Figure 3.1
Early ERO standardization of representations of eugenic traits.
Source: Charles Davenport et al., *The Study of Human Heredity* (1911), 4.

to pick out more specific conditions), color coding, and indicative arrows, arriving at representations of the family genealogy of specific eugenic traits, such as that in figure 3.2.

This kind of genealogical charting quickly became part of the standard representation of eugenic traits, particularly in work conducted or supported by the Eugenics Records Office. Its rapid adoption here rested on the credibility that the ERO had already established in research central to three other emerging fragile sciences on which eugenics drew: systematics, genetics, and anthropology. By the early twentieth century, tree structures had formed part of the standard representation of Darwin's theory of common descent, adapting a longer-standing use of such representations in pre- and non-evolutionary frameworks in systematics. And once the terms "gene" and "genetics," coined in 1906 and 1909, respectively, gained acceptance as the discrete, hereditary units that were transmitted from parents to offspring, pedigree charting of phenotypes became a proxy for the corresponding chart of alleles, the specific forms that the genotype might take at a locus.

In the study of kinship in anthropology, the publication of W. H. R. Rivers's "The Genealogical Method of Anthropological Inquiry" in 1910 is often regarded as consolidating (and occasionally as introducing) such pedigree charts in kinship studies, becoming a standard way to represent kinship relations. Yet the standardization of "the genealogical method" in the study of kinship crystallized gradually over a fifty-year period, with its methodological origins as a way to collect and represent ethnographic data

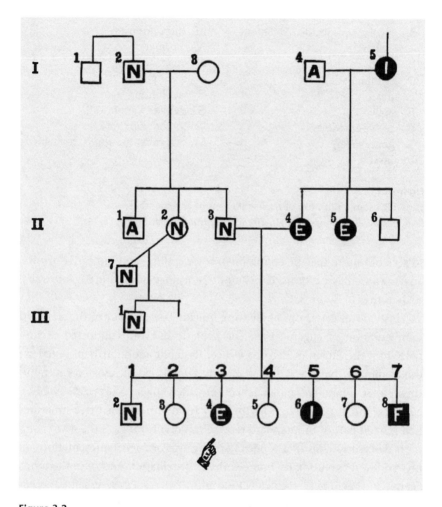

Figure 3.2
A simple chart of eugenic traits. In this chart, we see three generations represented,
containing family members who were putatively alcoholic (A), normal (N), epileptic
(E), insane (I) and feeble-minded (F).
Source: Charles Davenport et al., *The Study of Human Heredity* (1911), 12.

about kinship tracing all the way back to the beginnings of kinship theory. One of anthropology's founding figures, Lewis Henry Morgan, had introduced such modeling with his schedules in collecting data from the Iroquois in the 1850s, with their standardized use being extended to Melanesia and Australia in the 1860s. As the anthropologist Patrick McConvell and the historian Helen Gardner have recently shown, it was in this early extension from northern to southern studies in kinship that the representational innovation of tree construction was collaboratively innovated within the new fragile science of anthropology by Alfred Howitt and Tulaba [Toolabar] in Gippsland in Australia in the 1870s.

Davenport's standardization and systematization of pedigree charting in eugenics was not the only early contribution that the Eugenics Records Office made to the study of eugenic traits. Alongside the methodological works that offered this kind of standardization and systematization, the ERO also provided financial and mentoring support for an extension of the eugenic family studies themselves, as well as the extensive fieldwork that lay behind them and other research work undertaken by ERO staff and their allies. Seven of the ten eugenic family studies that were published between 1912 and 1926 were produced by ERO staff or trainees: Florence Danielson and Charles Davenport's *The Hill Folk: Report on a Rural Community of Hereditary Defectives*, and Arthur Estabrook and Charles Davenport's *The Nam Family: A Study of Cacogenics*—both published in 1912; Mary Storer Kostir's *The Family of Sam Sixty*, Anna Wendt Finlayson's *The Dack Family: A Study in Hereditary Lack of Emotional Control*, and Estabrook's *The Jukes in 1915*—all published in 1916; Mina Session's *The Feeble-Minded in a Rural County of Ohio* (1918), and Arthur Estabrook and Ivan McDougle's *Mongrel Virginians: The Win Tribe* (1926), published shortly before the famous U.S. Supreme Court decision in the case of *Buck v. Bell*.

Two other family studies—Henry H. Goddard's *The Kallikak Family: A Study in the Heredity of Feeble-Mindedness* (1912) and Elizabeth Kite's "The Pineys" (1913)—as well as Kite's "Two Brothers" (1912), which reflects the work completed by Kite that informed *The Kallikak Family*, derive from research undertaken at the Training School at Vineland, of which Goddard was director and by which Kite was employed as a fieldworker. The remaining work in the series, Arthur Rogers and Maud Merrill's *Dwellers in the Vale of Siddum* (1919), was also produced by a training school director and a coworker; Rogers was the superintendent of the Minnesota School for Feeble-Minded at Fairbault for more than thirty years.

Yet it was not only through such published work that the ERO articulated a certain conception of eugenic traits. Starting shortly after its founding, the ERO also sought to build a sweeping analytic index of traits in the U.S. population. By 1918 the ERO had amassed over half a million index cards charting the flow of eugenic traits through families, concentrating on traits regarded as deleterious or socially problematic, and approximately one million such cards by the time of the closure of the ERO in 1939. In addition to the mental and social traits that were to be found in eugenic laws, this research also investigated the genealogy of putative personality traits (e.g., rebelliousness, liveliness, and nomadism) as well as physical and physiological traits (e.g., diabetes, stature, and color blindness). As the historian of eugenics Garland Allen has pointed out, the ERO organized what otherwise might look like a miscellaneous list of traits into five categories: physical traits (stature, weight, eye and hair color, deformities); physiological traits (biochemical deficiencies, color blindness, diabetes); mental traits (intelligence, feeble-mindedness, insanity, manic depression); posited personality traits (liveliness, morbundity, lack of foresight, rebelliousness, irritability, missile-throwing, popularity); and social traits (criminality, inherited scholarship, alcoholism, patriotism).

"What is a trait?" is a serious and difficult question in the philosophy of biology, just as the more general "What is a property?" is a serious and difficult question in metaphysics. But we don't need to get too serious (and perhaps into philosophical difficulty) by taking up such questions in order to recognize that there is a problem with the expanding study of "eugenic traits." As the motley list of these traits themselves suggests—particularly those classed as mental, personality, or social traits—being identified as a eugenic trait in this research program is no guarantee of being a trait at all. At least not if we think of a trait minimally, in this context, as some kind of enduring and underlying property that is the cause of socially undesirable outcomes.

That problem to one side, what is most striking about the concept of a eugenic trait that emerges from the efforts of the ERO is just how recklessly sweeping the scope of eugenics was thought to be among those at the forefront of organizing and coordinating eugenic research and public outreach in at least North America. The idea that a eugenic trait is whatever those at the Eugenics Records Office collected data on gives us some sociological insight into how the notion of a eugenic trait functioned in practice. Yet in their aim to be comprehensive in their documentation of the potential reach of eugenics as a science, those directing eugenics research at the ERO seem to have forgotten about the causal and explanatory role that eugenic traits were meant to play in that science. Those traits were, after all, supposed to be keys to unlocking a rosy eugenic future, transmissible features that stemmed from something deep inside individuals that cause their defectiveness and social inadequacy. But in ranging from Huntington's chorea (or disease) to "seafaringness," the study of eugenic traits under the direction of Charles Davenport at the ERO undermined much of the credibility of eugenics as science. Thus, although such research publications provide us with much valuable information about this eugenic vision, it is less useful in providing an answer to the question of what a eugenic trait really is, unless we remain satisfied with an answer that encompasses the motley of traits cataloged by the ERO.[2]

3.3 Popular Culture

It was in the nature of the eugenics movement in general, and a function of the ERO's mission in particular, to blur the line between research and popular culture. In media coverage and popular cultural representation, advocacy for eugenics was often in terms of the threat posed to present and future society by the following sorts of people: the unfit, the feeble-minded, defectives, criminals, paupers, and prostitutes. Some of these traits, such as feeble-mindedness, are explicitly dispositional. That is, they are intrinsic tendencies of a person that are responsible for a range of behaviors or actions. Other traits, such as criminality, a person possesses because she engages in certain behaviors or actions. In general, however, the latter eugenic traits were typically thought of as at least implicitly dispositional. Criminals, prostitutes, and paupers, for example, were themselves thought to be sorts of people with certain inherent tendencies. This eugenic pull to the intrinsic is one form of individualistic bias in the fragile sciences with deeper political ramifications than that of other forms of such bias.

Since early in the short history of eugenics, eugenic themes have found their way into print media for popular consumption (books, magazines, newspapers, comics), novels and plays for middlebrow audiences, and paintings, movies, video, and video games for consumers of visual culture. This process of diffusion continues to this day. The genre of science fiction has been especially influential here, with human and other agents shaped by technology and social policy being a theme in late nineteenth-century novels, such as Robert Louis Stevenson's *The Strange Case of Dr. Jekyll and Mr. Hyde* (1886) and H. G. Wells's *The Island of Dr. Moreau* (1896). (Margaret Atwood's recent trilogy *Oryx and Crake* [2003], *The Year of the Flood* [2009], and *Maddaddam* [2013] represents the theme's newgenic update in contemporary fiction.) In the medium of film, eugenic movies shifted from more explicitly pro-eugenic propagandistic films, such as *The Black Stork* (1917) and *Tomorrow's Children: The Goals of Eugenics* (1934) to more narratively sophisticated films, perhaps the best known of which in contemporary culture is *Gattaca* (1997). As this small sampling of examples suggests, eugenic themes in popular culture are mixed with cross-cutting themes concerning human variation, genetic modification, transhumanism, political domination, environmentalism, utopianism, and dystopianism.

Like research publications emanating within the eugenics movement, popular culture is also limited as a source of evidence for answering the question "What is a eugenic trait?" This is not because, like those working at the ERO (for example), contributors to popular culture included almost anything as a eugenic trait. Rather, it is because they have been typically focused on exploring the potential and putative outcomes or eugenic policies, or their absence, where these are entwined with a host of other social issues. As such, discussions of eugenics in popular culture tend to be based on "what if" hypotheses about individual traits and the policies they are governed by, rather than focused on those traits themselves. Representations of eugenics in popular culture are thus important for understanding the continuing infusion of eugenic ideas and practices into our conceptions of what is and is not possible in society, and are especially powerful as ways of motivating public engagement and discussion. But they are less useful for us in the task at hand here.[3]

3.4   Eugenic Laws: Marriage and Immigration

While eugenic thinking was most directly expressed in sexual sterilization legislation, it also found its way into other kinds of provincial, state, and federal laws and policies in North America. The two most notable examples were marriage restriction laws and immigration laws and policies. Like other laws and policies that aim to improve human society, laws regulating marriage and immigration could in principle operate by facilitating marriage and immigration for all. But in keeping with the dichotomizing nature of The Eugenic Mind, those that do so in the name of eugenics have differentiated between different sorts of people, promoting the marriage and immigration of some, while discouraging or even prohibiting that of others. By restricting who can marry, one inhibits the reproduction of some sorts of people, and by facilitating the immigration of some sorts of people and inhibiting that of others, one differentially impacts what future populations look like—sometimes literally. Both marriage restriction laws and immigration laws have functioned as part of explicitly nation-building policies to shape future generations in Canada, the United States, and elsewhere, and they have been consistently (even if not constantly) directed at excluding or eliminating the sorts of people identified as posing a eugenic threat to the health of the nation.

The eugenic uses of both marriage restriction and immigration laws have usually been viewed as focusing primarily on race and ethnicity. Restrictive laws centered on race have predominantly taken the form of anti-miscegenation laws in the United States, while those articulated in terms of ethnicity, as we saw in chapter 2, have primarily operated via quota-based or point-driven criteria for immigration that place a special burden on people from certain ethnic backgrounds in both Canada and the United States. Less frequently noted is that both kinds of laws were often designed to impose an intergenerational bottleneck on the reproduction of people with disabilities. For example, marriage restriction laws forbidding or limited the marriage of people with disabilities were in place in more than thirty states in the United States. Canadian immigration law, starting with the 1869 Act Respecting Immigration and Immigrants through to the 2001 Immigration and Refugee Protection Act has a long history of restricting the immigration of those with disabilities, including the contemporary appeal to the "undue burden" clause of current legislation used to both deny admission to Canada and to deport people already living in the country.

The best known of immigration laws in North America directed by eugenic concerns is the Johnson-Reid Act in the United States, signed into law by President Calvin Coolidge in 1924, and mentioned briefly in chapter 2. That law placed a quota on the number of immigrants from Asian and Southern and Eastern European countries, capping these numbers at 2 percent of the 1890 U.S. Census population who had originally emigrated from these countries. That law was aimed to reverse "the passing of the great race" and was influenced by Madison Grant's book of that name. The law was also shaped by the appointment of the Eugenics Record Office's Harry Laughlin as an advising "eugenics expert" from 1921. Laughlin's long-standing view that immigration had become a major source of degeneracy and a dysgenic influence on the American population played an important role in guiding legislators in the articulation and passing of the Johnson-Reid Act.

But the reach of eugenics into immigration policy went well beyond that act and the borders of the United States. For example, following the 1902 Royal Commission into Chinese immigration, the Canadian Immigration Act of 1910 barred the entry of "immigrants belonging to any race deemed unsuited to the climate or requirements from Canada," while the 1923 "An Act Respecting Chinese Immigration" severely limited the immigration into Canada of "persons of Chinese origin or descent irrespective of allegiance or citizenship."

While marriage and immigration restriction laws no doubt shed light on key aspects of eugenics, one of the limitations of these as sources of evidence for answering the question of what eugenic traits are is that such laws were the result, and often predominantly so, of ideological factors other than eugenics, such as racism. For example, the first anti-miscegenation law in North America originated in 1664 and banned the marriage of whites and slaves, with forty-one of the fifty states in the United States enacting anti-miscegenation laws over the next three hundred years before they were deemed unconstitutional by the U.S. Supreme Court in *Loving v. Virginia* in 1967. Doctrines of racial purity and the racism they both resulted from and fueled in turn came to stand in two-way relationships with eugenics. But since marriage restriction laws are the result of this longer reach of racism, they present at best a partial guide to what is properly considered a eugenic trait. The same general point holds true, I want to suggest, of immigration restriction laws, even when they are clearly motivated, as was the Johnson-Reid Act, by explicit eugenic concerns. They too have a history that extends beyond the short history of eugenics that limits their value as indicators of eugenic traits.[4]

3.5 Sexual Sterilization Legislation

As laws introduced specifically and self-consciously to combat a perceived eugenic threat, sexual sterilization laws do not suffer from this problem. These explicitly eugenic laws—as manifest in their titles and the arguments given for introducing them—are significantly less mediated manifestations of eugenic thinking than are marriage and immigration restriction laws. For this reason, I want to provide a more detailed analysis of what eugenic traits one finds as grounds for sterilization in order to reveal more about just what eugenics traits are, and then turn to how eugenics should be understood in light of that analysis.

Two provinces in Canada and thirty-three states in the United States passed eugenic sterilization laws, starting in 1905 with the state of Pennsylvania, whose original sterilization law was passed by the state legislature but was vetoed by Governor Pennypacker. Since most jurisdictions revisited their initial enactments, there were 77 statutes or amendments of statutes pertaining directly to eugenic sterilization legislation in North America in the twentieth century, including 7 that were either vetoed by governors or rescinded by referendum. Across these 77 statutes, there were approximately 300 eugenic trait instances; if we remove those instances that recur across a state's or province's laws, we have 180 trait instances. These specified approximately twenty grounds for eugenic sterilization (see table 3.1).

As table 3.1 clearly shows, the dominant cluster of traits in eugenic sterilization legislation concern psychological traits and mental health— feeble-mindedness, insanity, imbecility, idiocy, mentally unfit/deficient, mental disease/illness, psychotic—accounting for 165 of the approximately 300 instances of all traits mentioned, and 105 of the 180 instances of all traits when we restrict this count by avoiding repeats within a jurisdiction in later legislation. In either case, the cluster of psychological and mental health traits subsume over 55 percent of all eugenic traits. If epilepsy is added to this mental health cluster, which arguably it should be, then these numbers move to 204 of 300, and 132 of 180, approximately 70 percent of eugenic traits mentioned in sterilization laws. Thus, such laws clearly reflect a focused eugenic concern about psychological traits and mental health, whether one includes jurisdictional amendments to existing legislation, and whether one includes epilepsy. Of these, "feeble-mindedness"

Table 3.1

Eugenic traits in North American eugenic sterilization laws

Trait/ground	Statute frequency (total = 77)	Frequency by state/ province (total = 35)
Feeble-mindedness	48	29
Insanity	43	27
Epilepsy	39	27
Criminality	32	17
Imbecility	25	17
Idiocy	23	17
Sexual perversion/depravity	22	11
Mentally unfit/deficient	14	7
Moral depravity/degeneracy	16	8
Rape	8	7
Mental disease/illness	8	5
Syphilis	6	4
Institutionalized person	7	6
Pedophilia	2	2

Note: Data compiled by Luke Kersten and the author from the following sources: Kaelber, *Eugenics* (n.d.); Landman, *Human Sterilization* (1932); Laughlin, *Eugenical Sterilization in the United States* (1922); Lombardo, *Three Generations, No Imbeciles* (2008); Paul, "*... Three Generations of Imbeciles Are Enough ...*" (1965); and copies of the legislation itself. The table omits the following traits, which occur only in single jurisdictions: alcoholism (Iowa 1913), moron (New Jersey 1913), incestuousness (Nebraska 1929), Huntington's chorea (Alberta 1942), psychotic (Alberta 1937, 1942).

appears as a eugenic trait in 29 of 35 jurisdictions; the exceptions are the Canadian provinces of Alberta and British Columbia, and the U.S. states of Alabama, Nevada, Utah, and Georgia.

Apart from the cluster focused on psychological traits and mental health, there is a smaller, secondary cluster of eugenics traits that concern sexual behavior—sexual perversion/depravity, moral depravity, rape, pedophilia, incest—accounting for 49 of the 300 trait instances (29 of the 180), or around 15 percent of the total. If one adds the trait *criminality* here, which more arguably than in the previous case of epilepsy one should, this behavioral cluster accounts for approximately 25 percent of all eugenic traits, again whether one includes or excludes jurisdictional amendments to existing legislation.

This very basic analysis highlights the predominant place that psychological traits and mental health had in the way in which the eugenic intervention of sexual sterilization was conceptualized and justified, reinforcing the view of eugenics that I sketched in chapter 2. Psychology and mental deficiencies posed a particular threat to the intergenerational fitness of the population, including (and most specifically) the so-called "menace of the feeble-minded." But this crude analysis also serves as a quantitative reminder that a significant, secondary dimension to eugenic legislation in North America was directed at those who violated social and legal norms, particularly those related to the overlapping categories of sexuality and crime. Sterilization based on these eugenic traits tended to be more punitive in nature: either a form of punishment or simply a means to ensure that the crime, whether actual or imagined, would not happen again.[5]

3.6   Alberta at the Legislative Margins

There is, no doubt, more that could be said about these clusters and of the eugenics traits within each of them. Rather than elaborate on the mainstream trends in thinking about eugenic traits, however, I want to focus instead on what can be said about the legislative margins that lie beyond the clustering and such trends. More particularly, in this section I shall concentrate on a way in which the original sexual sterilization legislation in Alberta from 1928 and its 1937 and 1942 amendments were distinctive. In the following sections, I turn to say what such a consideration of eugenic traits at the legislative margins tells us about eugenic traits and the social mechanics of eugenics.

The first thing that is striking about Alberta's original Sexual Sterilization Act is that it did not refer to any *real* eugenic traits as grounds for sterilization. Alberta was not unique in this respect but found itself in the company of a small number of U.S. states: Indiana, California, Alabama, North Carolina, and Georgia. All six North American jurisdictions had eugenic sterilization legislation that was not articulated in terms of one or more eugenic traits that a person might possess. Rather, their legislation was cast instead simply in terms of an institutionalized person. Once someone was an inmate, resident, prisoner, or patient at an appropriately designated institution, they were, simply in virtue of that fact, eligible for sterilization, following some kind of process of review or examination. Unlike the comparable preceding legislation in Indiana (1907), California (1909), Alabama (1919), and North Carolina (1919), the original legislation in Alberta did not apply to those in prisons, but only to persons institutionalized at what in Alberta was referred to as a "mental hospital." The crucial criterial clause of the original Alberta legislation is striking, and unique. It contains one long sentence, and reads in full:

If upon such examination [by the "board of examiners," i.e., the Eugenics Board], the board is unanimously of opinion that the patient might safely be discharged if the danger of procreation with its attendant risk of multiplication of the evil by transmission of the disability to progeny were eliminated, the board may direct in writing such surgical operation for sexual sterilization of the inmate as may be specified in the written direction and shall appoint some competent surgeon to perform the operation.

The striking phrase "risk of multiplication of the evil by transmission of the disability to progeny" is not only unique, but so too are its components. This is one of the very few places in which the words "disability" and "progeny" are explicitly used in sexual sterilization legislation, and the only place where there is talk of "multiplication of the evil" of the transmission of any characteristic.

In 1937 Alberta amended its legislation explicitly to solve a pair of problems that had arisen in the eyes of pro-eugenicists. The first of these reflected the concern that many of those who should be sterilized were not housed in a "mental hospital" but were living in the broader community. The second problem concerned consent. The 1928 legislation required that consent be obtained to proceed from approval to sterilization, either from the "inmate," where that person is deemed capable of giving consent, or his or her spouse or parent, otherwise. This original consent requirement had led to sterilization board approvals outstripping sterilizations, since such consent was often withheld or not given. Together with the belief that there remained many targets of eugenics who simply lived within the general population, this fueled the view that the Sexual Sterilization Act required an expansion on two fronts if it were to be more effective.

The amended legislation addressed both problems. First, it expanded the range of recommending institutions from "mental hospitals" to include what were called "mental hygiene clinics." These clinics, sometimes referred to as *guidance clinics*, had started up in Alberta shortly after the original Sexual Sterilization Act had been passed, and they took the form of mobile, traveling clinics, visiting schools and public health offices that offered assessments and evaluations of the mental abilities and proclivities of students and patients. Second, the amended legislation introduced and distinguished between a pair of eugenic traits—between *psychotic persons* and *mental defectives*—and required consent only for the former of these. These were defined in the amended legislation itself in section 2 of the amended law as follows:

(2c) "Mentally defective person" means any person in whom there is a condition of arrested or incomplete development of mind existing before the age of eighteen years, whether arising from inherent causes or induced by disease or injury.
(2e) "Psychotic person" means a person who suffers from a psychosis.

The corresponding eugenic traits map roughly onto what we would now refer to, respectively, as developmental intellectual disabilities, on the one hand, and schizophrenia and other delusional psychiatric conditions that are manifest typically beyond childhood, on the other. Both definitions are quite sweeping. The first explicitly includes those whose arrested or incomplete development is the effect of life's experiences. Such experiences could be incredibly diverse and varied; the relevant resulting conditions include injuries acquired through accidents or malice and cognitive limitations that arise from infectious diseases, such as syphilis or tuberculosis.

In effect, this amendment increased the already considerable power wielded by recommending superintendents and the approving Eugenics Board. There were now two eugenic traits in play, being *psychotic* and being *mentally defective*, both characterized legislatively in terms of an individual being a certain sort of person, a psychotic or a mental defective. While being institutionalized in a "mental hospital" remained the only criterion that such people needed to satisfy to be recommended for sterilization by the institution's medical superintendent, being mentally defective became criterial for being recommended for sterilization for those seen at a mental hygiene clinic. Since children were more likely to be "mentally defective persons" than "psychotic persons," and schools became a major site for eugenic intervention via those clinics, the effect here was to more effectively identify children as targets of eugenics.

The other chief effect of the introduction of this pair of eugenic traits concerned not eugenic intake but post-recommendation outcomes. This effect interacted with the heightened focus on children. The relaxation of the consent requirement meant that approval by the Eugenics Board for the sterilization of someone deemed to be mentally defective made sterilization itself highly probable. And children and others in the community who had not been institutionalized, but whom a medical practitioner at a guidance clinic had deemed mentally defective could be required "to be examined by or in the presence of the Board." With this amendment, the eugenic pipeline in Alberta was increased in length and diameter, extending in reach and volume the province's practice of eugenics.

The amended 1937 legislation in Alberta was the only piece of legislation in North America to mention psychotic persons (or psychosis) as a eugenic trait. In using the language of mental defectives rather than "the feeble-minded," Alberta was also unique in having an original and continuing focus on mental health without mentioning feeble-mindedness explicitly as a eugenic trait. The trajectory of distinctness here was continued in the 1942 amendment to the legislation, which added neurosyphilis, Huntington's chorea, and epilepsy to the list of eugenic traits. Neither neurosyphilis nor Huntington's chorea appear as eugenic traits in any other piece of legislation in North America, and the addition of epilepsy as a eugenic trait after 1930 occurs only in two other jurisdictions, Oklahoma (1931) and South Carolina (1935). The other eugenic trait here, epilepsy, did feature in approximately 75 percent of North American jurisdictions with eugenic sterilization legislation. Yet by 1942, just when it is added as a eugenic trait in Alberta, epilepsy had come to be viewed in general as a dubious eugenic ground for sterilization. Indeed, the addition of epilepsy as a eugenic trait after 1930 occurs only in two other jurisdictions, Oklahoma (1931) and South Carolina (1935). I shall return in concluding this chapter with some closing thoughts on each of these three traits and the social mechanics of eugenics in Alberta.

The 1942 amendment that added these three eugenic traits as the basis for sexual sterilization in Alberta required consent of the person recommended for sterilization. As such, those newly subhumanized in the amendment were assimilated to the category of "psychotic persons," rather than "mental defectives," following the division introduced in the 1937 amendment. Given that that bifurcation roughly tracked that between children with intellectual disabilities and adults with certain psychiatric conditions, the 1942 amendment in effect expanded the range of "psychotic persons." Unlike the 1937 amendment, which aimed primarily to increase the sterilization rate of children deemed to be mentally defective, the 1942 amendment represented a relatively minor fine-tuning of the provincial effort to sterilize adults who were viewed as psychiatrically compromised.

By the time Alberta had introduced the Sexual Sterilization Act in 1928, over 80 percent of the sterilization legislation there would be in North America had been introduced, repealed, and amended. By the time of the 1937 amendment to that act that aimed to extend sexual sterilization practices directed at children in the province, the practice of sexual sterilization on eugenic grounds had come under increasing scrutiny in the rest of North America. By the time of the 1942 amendment, the eugenic traits that Alberta was adding to its legislation—neurosyphilis, Huntington's chorea, and epilepsy—were seldom regarded as eugenic traits in other such jurisdictions. And by the time Alberta repealed its legislation in 1972 after twenty-five years of vigorous postwar implementation, eugenic sterilization had already been viewed throughout this period as a matter of a dark past in much of the world. Certainly within the North American eugenics movement, Alberta lay very much at the legislative margins. Yet there is sometimes general knowledge to be gained at the margins.[6]

3.7   Institutionalization and the Social Mechanics of Eugenics

One of the primary ways in which policies of sexual sterilization were put into practice was the prior segregation of people through the process of institutionalization, typically in an asylum, home, hospital, or training school for mental defectives, the feeble-minded, or the insane. The overwhelming majority of legislation explicitly specified some further eugenic trait, typically thought of either as some kind of intrinsic property of the person—such as is the case for all traits in the psychological and mental health trait cluster—or as a function of behavior that the person has engaged in—such as is the case for all of the traits in the sexuality/criminality cluster. But as mentioned in passing in section 3.6, in just six jurisdictions—Indiana (1907), California (1909), Alabama (1919, 1934), North Carolina (1919), Alberta (1928), and Georgia (1935, 1937)—the legislation allowed or mandated sterilization for any person so institutionalized. In these cases, being an institutionalized person itself functioned as a eugenic trait, not simply as a conduit for being sterilized on the basis of putatively having some other trait. The relationship between having been institutionalized and sterilization functioned very differently, however, in the two national contexts.

Within the United States, serious concerns about the constitutionality of the original laws in Indiana, Alabama, and North Carolina led to extremely limited use of the original laws in those jurisdictions. Most sterilizations from Indiana seem to have been done outside of the terms of the law there; there were no reported sterilizations in Alabama up until 1927, and only forty-four up until 1929; and in North Carolina there appears to have been no sterilizations under the 1919 law. Likewise, the 1935 law in Georgia was vetoed by Governor Eugene Talmadge, and the 1937 law that was passed only led to sizeable numbers of sterilizations from 1941.

By contrast, this constraint in practice was absent in the Canadian province of Alberta. Instead, the number of sterilizations was limited chiefly by the requirement to get consent, a requirement that, as we have noted, was significantly weakened in the first amendment to the Sexual Sterilization Act of Alberta, passed in 1937. As we have noted, this amendment distinguished between "mentally defective" and "psychotic" persons, requiring consent only for the sterilization of the latter. That legislation also broadened the institutional basis for sterilization, moving from "any inmate of a mental hospital" to "any patient of a Mental Hospital" *and* "any mentally defective person who has been under treatment or observation at a [mental hygiene] clinic." Such clinics had been introduced in Alberta in 1929, functioning in their first eight years as traveling clinics that served in part to feed institutions, such as the Provincial Training School. With the 1937 amendment, the Sexual Sterilization Act made it possible for those deemed mentally defective at a guidance clinic to appear before the province's Eugenics Board without having to be admitted to such institutions.

In terms of this general transborder difference, California was an exception. Its 1909 law, which authorized sterilizations in the California Home for the Care and Training of Feeble-Minded Children, in any state hospital, *and* in any state prison, was repealed in 1913 when an even more encompassing sterilization law was passed that specified hereditary insanity, incurable chronic mania or dementia, and idiocy as eugenic traits. In a

1917 law there was a further broadening of those traits to encompass those "afflicted with mental disease which may have been inherited and is likely to be transmitted to descendants, the various grades of feeble-mindedness, those suffering from perversion or marked departures from normal mentality or from disease of a syphilitic nature."

What five of these six jurisdictions shared, however, were extensive and long-lasting sterilization practices, once they got started. The exception here is Alabama, whose 1934 legislation, which sought to extend the range of persons the law applied to, was both vetoed by Governor Bill Graves in that year and also deemed unconstitutional by the Alabama Supreme Court in 1935. In effect, this halted the practice of state-mandated sterilization altogether in Alabama. The earliest states to sterilize—Indiana and California—both increased their rates of sterilization after subsequent legislation was introduced, respectively, in 1927 and 1917, and went on to sterilize at high rates, relative to other jurisdictions, over subsequent thirty-year periods. The other three jurisdictions—North Carolina, Georgia, and Alberta—began more extensive practices of sterilization in 1937–1938, and had three of the highest five rates of sterilization (together with Virginia and California), sterilizing, respectively, approximately 7,600, 3,300, and 2,800 people in the ensuing thirty-five-year period. Unlike Virginia and California, they did so chiefly after 1945. While all three of these jurisdictions passed amended laws that referenced the eugenic trait of mental deficiency or defectiveness in 1937–1938, their preexisting practices of sterilization directed at "inmates" or "patients" more generally had already functioned to give individuals with specific roles—superintendents and members of eugenics boards—a free hand in the recommendation and approval of sterilization. The same would seem to be true of Indiana (which sterilized 100 people *before it became the first state to enforce sexual sterilization legislation*), and California, which sterilized one-third of all those sterilized in the United States.

We can consider several hypotheses as to why there appears to be a strong correlation between appealing solely to someone's being an institutionalized person as the grounds for subjecting them to sexual sterilization and engaging in extensive and long-lasting sterilization practices. First, passing such legislation through democratic means shows a kind of confidence in the reliability of institutionalization as a means to identifying only people deemed to warrant sexual sterilization: there was no need to specify distinct eugenic traits of the person because of the faith in that reliability. Second, in the absence of such eugenic traits, the superintendents from those institutions making recommendations of who to sterilize and the eugenic boards considering such recommendations were given tremendous power in deciding the fates of those over whom they exercised the relevant authority. This not only reflects the legislative bravado shown in the passing of such legislation, but also made a clear difference to the practice of eugenics itself. And as was the case in at least one of those jurisdictions—Alberta—the hubris shown here had disastrous consequences for people who were subsequently sterilized, many of whom were wrongfully institutionalized in the first place.[7]

3.8   Mental Defectives and the Mentally Ill: Beyond Consent

Despite uniquely using "psychotic person" as a term to refer to a eugenic trait, Alberta's amendment to the Sexual Sterilization Act in 1937 introducing the distinction between mental defectives and psychotic persons mirrored the widely recognized bifurcation between the two major classes of eugenic traits that were focused on mental health. As we saw in chapter 2, this evolving distinction became a more entrenched dichotomy between the feeble-minded (mental defectives, idiots, imbeciles), on the one hand, and the insane (the mentally ill, lunatics), on the other hand.

As we have also noted, mental deficiency is typically taken to be what we would now call a *developmental disorder* or *disability*. As such, it was often detected in or ascribed to children. The establishment of the Provincial Training School for Mental Defectives in 1923 in Red Deer, Alberta, as well as the introduction of guidance clinics in 1929 that often operated through the public education system, served to detect, diagnose, and treat children deemed to be mentally defective. With the introduction of the Sexual Sterilization Act in 1928, this treatment included sterilization; with the amendment to the Act in 1937 that removed the requirement of consent for mental defectives, this increased the sterilization rates among children in particular.

Yet as the testimony of eugenics survivors like Leilani Muir, Judy Lytton, Glenn George Sinclair, Roy Skoreyko, and Ken Nelson make clear, it was not simply that consent was not legally required. Children were sterilized typically even without their knowledge of what was being done to them, and what this might mean for their pathways in life. In some cases, their ignorance about these life-lasting surgical interventions extended into their adult lives, adding the psychological trauma of deception to the wrongfulness of what had been done to their bodies. This dimension to the effects of the institutional complicity of the Provincial Training School and the Alberta Eugenics Board remains a largely unacknowledged form of dehumanization of those made into eugenic targets.

In at least Alberta, children deemed mentally defective, or feeble-minded, were positioned beyond consent. And beyond the consent that was no longer needed to sterilize them, for the thirty-five years following the 1937 amendment, unknowing children became unknowing adults whose life courses were shaped, often dramatically, by their unwillful ignorance about their own bodily state. How widespread this targeting of children was in North American sterilization practices more generally remains a topic in need of further research, as do the effects of coming to terms with a wrongfulness unacknowledged and hidden from view for much of one's adult life.[8]

3.9 Three Eugenic Traits: Syphilis, Huntington's, and Epilepsy

In section 3.6 I made some brief, general comments about the 1942 addition of neurosyphilis, Huntington's chorea, and epilepsy to the list of eugenic traits specified in the Sexual Sterilization Act of Alberta. In sections 3.7 and 3.8 I drew some general conclusions about the social mechanics of eugenics vis-à-vis institutionalization and consent based on the perspective afforded by the legislative margins occupied by Alberta in North American eugenics. To conclude the chapter, I return to focus on those three late additions to the list of eugenic traits: syphilis, Huntington's chorea, and epilepsy.

Contagious diseases such as syphilis and tuberculosis are occasionally mentioned in more general discussions of eugenics, both past and present. For example, speaking of marriage restriction laws in their introduction to *The Oxford Handbook of the History of Eugenics,* Philippa Levine and Alison Bashford say "there were numerous attempts by eugenic associations to make marriage screening compulsory, aiming to restrict the reproduction of those with conditions and diseases considered heritable: syphilis, leprosy, tuberculosis, epilepsy, alcoholism, and less specific conditions such as 'criminality' or sexual 'tendencies.'"

While Levine and Bashford are correct in implying that people with syphilis were sometimes the targets of eugenics, there is less evidence that it (along with leprosy and tuberculosis) was generally conceived as a *heritable* disease or condition. Contagion and heritability represent independent modes of transmission, including intergenerational transmission, of undesirable characteristics, and The Eugenic Mind sometimes ranged over both. However syphilis was thought of in popular culture, prior to 1942 it appeared as a eugenic trait in original sterilization legislation only in Iowa in 1911, and in amended legislation only in California in 1917. (South Dakota added it to existing eugenic legislation in 1943.) In fact, it appears that very few people were sterilized on the basis of having syphilis, despite both California and Iowa being states that had high rates of sterilization. The focus of sterilization in such states was very much on, respectively, the mentally ill and the feeble-minded.

While syphilis is thus a seldom-invoked eugenic trait, the term "neurosyphilis" is rarer still, being almost never mentioned in relation to eugenics at all, and with Alberta's 1942 legislation providing the only instance where it functions as a eugenic trait guiding a sterilization mandate. One of the few places in which it is used in connection with eugenics is in a 1934 government report, authored by Provost John MacEachran, entitled "Social Legislation in the Province of Alberta," which includes a broad overview of health services in the province. In the section of the report on "Venereal Disease Control," MacEachran mentions the "early detection of neurosyphilis" and its prevention as being especially important for mental illness, immediately before he then turns to report on mental hygiene in the province. It would thus not be surprising if at least the terminological idiosyncrasy derives directly from MacEachran's influence, one that incorporated a contagious disease as a eugenic trait because of its putative relationship to mental illness.

Huntington's chorea or disease, the second of three eugenic traits introduced in Alberta's 1942 amendment to the Sexual Sterilization Act, is most closely associated with mental deterioration in the second half of life, and is likewise absent from all other sexual sterilization legislation in North America. Charles Davenport had recommended that Huntington's be grounds for sterilization in early publications of the Eugenics Record Office. But despite this, and the fact that Huntington's was known at the time to be one of the few Mendelian autosomal dominant traits, requiring the presence of just one copy of the allele for the disease to be manifest later in life (typically between thirty-five and forty-five years of age), this recommendation did not find its way into any North American legislation. Davenport's limited influence, in this respect, matched that of Henry Goddard's in Goddard's introduction of the eugenic trait *moron*. Despite the prevalence of "moron" in popular culture and eugenic propaganda, it was placed in the legislation of only one state. That state was New Jersey, the state in which Goddard's Vineland Training School for Feebleminded Boys and Girls was located.

Although epilepsy features as a eugenic trait in twenty-seven of thirty-five jurisdictions, I noted in section 3.8 that Alberta's inclusion of it in its 1942 amendment set it apart from all but two states, joining only Oklahoma (1931) and South Carolina (1935) as jurisdictions in which epilepsy was added as a eugenic trait after 1930. The relevance of this is that by the early to mid-1930s, there was sizeable opposition to practicing eugenic sterilization on those whose sole eugenic trait was being epileptic. A widely read report from a committee of the American Neurological Association from that time had noted that the "term has been dropped by many workers, who now speak of the convulsive disorders" and that the committee will "retain its use in this report merely because the heredity studies in the literature bear this name." Given that epilepsy was viewed increasingly as at best a dubious eugenic trait throughout the 1930s in the rest of North America, its introduction as such a trait in Alberta's 1942 amendment requires explanation.

Like Huntington's chorea, epilepsy had also been the subject of an early report by Davenport that discussed epilepsy as a eugenic trait, linking it together with a cluster of other traits, such as feeble-mindedness and neuroticism in a study that featured the kind of pedigree analysis for which the ERO became well-known, as we saw in chapter 2. And like both Huntington's and neurosyphilis, epilepsy's undesirable effects were viewed as being mediated through the mind: all three were viewed as eugenic traits that involved mental breakdown, collapse, or deterioration in adulthood. One admittedly speculative hypothesis for why both Huntington's and epilepsy appear, strangely, in the 1942 amendment to the Sexual Sterilization Act of Alberta, despite both the rise of scientific views that suggested caution here and the dampening of their perception as eugenic traits over time, was that proponents and advocates of these changes relied on these early reports from the Eugenics Records Office, rather than current states of affairs or contemporary scientific knowledge. Whether all three traits made their way into the not-quite-so short history of eugenics in Alberta through the agency of Provost John MacEachran, as it appears that at least neurosyphilis did, remains an open question.[9]

4 Subhumanizing the Targets of Eugenics

4.1 What Sorts of People Should There Be?

Back in the fall of 2008, I was involved in co-organizing a public dialogue, on "The Modern Pursuit of Human Perfection: Defining Who Is Worthy of Life." This dialogue formed part of a longer series of community events and was cosponsored by the Alberta Association for Community Living—now called Inclusion Alberta—and by an informal network I had recently assembled, the What Sorts Network. The name of the network was a shorthand gesture to the question "What sorts of people should there be?" That question—interestingly, in its singular form—was also the title of one of the earliest philosophical books exploring technologically facilitated human enhancement.

Written by the philosopher Jonathan Glover, *What Sort of People Should There Be?* focused on the ethics of genetic engineering. Much of the burgeoning literature on genetic knowledge, intervention, and human enhancement over the past thirty-five years, including Glover's more recent *Choosing Children*, uses the history of eugenics as a framework for ongoing ethical issues about reproduction, parenting, human nature, and the uses of scientific knowledge and technology to modify human beings. What kind of framework that history provides depends very much on one's standpoint.

The perspective of philosophical bioethicists regarding eugenics is neatly encapsulated in Nicolas Agar's recent, short encyclopedia article, "Eugenics, Old and Neoliberal Theories of." There Agar begins by acknowledging the "terrible history" of eugenics, going on to distinguish two views of the relevance of that history to contemporary discussions: "On one view, the story of eugenics serves principally as a cautionary tale. Education about eugenics tells practitioners of human genetic selection or modification what they should avoid. Another view denies that eugenics is intrinsically immoral. Once purged of its multiple moral and scientific errors, Galton's science of improving human stock can give rise to morally acceptable forms of eugenics."

We have already glimpsed the "cautionary tale" view of the short history of eugenics in our brief discussion in chapter 1 of concerns about the Human Genome Project and the interventionist genetic technologies it has generated. As Agar himself notes, what we might refer to as *sanitized eugenics*—eugenics "purged of its multiple moral and scientific errors"— has been both explored and endorsed by many contemporary philosophers working in bioethics, and is explicit in one of the most widely read books in bioethics in the past twenty years, *From Chance to Choice*. Such explorations and endorsements most prominently concern *liberal* eugenics, a cluster of forms of eugenics that emphasize individual choice, rather than state intervention, as the key mechanism for improving human lives across generations. We shall return to take up sanitized eugenics in part II.

Whether adopting the cautionary tale view or considering sanitized eugenics, for bioethicists and moral philosophers who turn their attention to the relevance of eugenics today, eugenics *is* past. For them, eugenics is a historical backdrop of one kind or another to contemporary bioethical decisions and dilemmas that have a certain kind of distance from the short history of eugenics. That this perspective on eugenics could not simply be taken for granted was something that I learned from my experience with the "Modern Pursuit" dialogue.

That experience also made it clear to me that bioethics and even philosophical ethics more generally could not be assumed to provide neutral, objective perspectives on questions like "What sorts of people should there be?" In taking up themes of life worthiness and subhumanization in this chapter, I shall return more explicitly to the ideas of institutional complicity and engaged individuality introduced in chapter 1, particularly the limits of engaged individuality within bioethics as a manifestation of the circumscription of our prosocial nature.[1]

4.2 The Pursuit of Human Perfection and Life-Worthiness

The "Modern Pursuit" dialogue belonged to a national series of public events that the disability advocacy organization the Canadian Association for Community Living held as part of its fiftieth anniversary celebrations under the general heading "A Matter of Diversity." The general theme of the public dialogue—debates over the impact of new and emerging technologies on the sorts of people there should be in future generations—was interesting and potentially provocative. But that theme was not all that distinctive in the context of awareness of the eugenic past. More distinctive were three features of the dialogue.

First, there was a central focus on, rather than a passing mention of, the ways in which the uses of those technologies impact the lives of people with disabilities in contemporary Canadian society. As a part of this, attention was drawn explicitly to ways in which bioethics and medical ethics themselves functioned in both the articulation of ideas and the translation of those ideas into practices that were detrimental to the lives of people with disabilities. This was one form of institutional complicity in the propagation of views and practices impactful on, and often antagonistic toward, people with disabilities.

Second, there was a working assumption that the continuities between the eugenics of the recent past and contemporary attitudes and policies toward people living with disability were substantial. As mentioned in chapter 1, with the Sexual Sterilization Act of Alberta having been repealed only in 1972, many sterilization survivors still live within our local community. And for many others living with disability in that community, the issues of institutionalization and eugenic sterilization were very much ongoing concerns. For them, sentences that began "if I had been born in the 1950s in Alberta, ..." or "if those eugenic policies were still in place today, ..." were no mere counterfactuals imagined to hold (in true philosopher style) only in some distant possible world. They were an expression of a felt anxiety that formed part of their day-to-day lives.

Third, and most important, the dialogue was structured around a panel discussion led primarily by people with the lived experience of parenting with disability, either their own or that of their child. Their experiences of pregnancy and parenthood—in a world in which they were the targets of practices of prenatal screening and policies of child removal—were governed by the fear and shame associated with intellectual disability, rather than the potential joys many of us can take for granted in the experiences of pregnancy and parenting. They had been medically advised and counseled in terms of a negative notion of risk and a list of uphill battles and insurmountable challenges, rather than in terms of the neutral concept of chances and the opportunity to shape meaningful lives for both themselves and their families. The stories conveyed by these community members were powerful in many ways. In concert with the other two features of the dialogue, these stories provided an invaluable perspective that was typically absent from the usual discussions initiated and led by professorial experts in bioethics and medical ethics.

Not that I knew all that much about the intricacies of even those academic discussions. After all, I had been trained as a philosopher working primarily on the mind, cognitive science, and the philosophy of biology, rather than as someone specializing in bioethics, disability studies, and political philosophy. I had learned to be comfortable with both Twin Earth and twin studies. Over the proceeding five years or so, I had become familiar with Alberta's eugenic past. But at that time I had no real experience with community-based advocacy around issues of disability. The relationship between eugenics and disability was something that neither my personal experience nor my academic background primed me to reflect on. I was not tuned to the real-life consequences of eugenic ideas about human improvement and human perfection for people currently living with at least certain kinds of disability.

I knew that the translation of those ideas into practices in the eugenic past led to the subhumanization of those institutionalized and sterilized, having been classified—to use terms then common—as mentally defective, feeble-minded, or of degenerative stock. They had been deemed unworthy of a full and complete human life. In some cases, this became literally a matter of life and death for people so classified; such people had been deemed unworthy of life altogether.

These were severe judgments about the mentally defective, with real-life consequences for the quality and duration of their lives. But perhaps we should think in other terms of the social judgment that was, in effect, passed about them. For were the putatively mentally defective, feeble-minded, and degenerate only deemed unworthy of a full and complete human life, they might simply have faced passive neglect and negligence, and thus fallen outside of societal circles of protection and security that the rest of us could rely on. Yet historically those deemed mentally defective have been subject to more active practices of subhumanization, including their segregation through institutionalization, invasions of their bodily integrity, and even deliberate efforts to terminate their lives.

It was not simply that they were deemed defective, but that their defectiveness posed some kind of ongoing threat to society, one that required perhaps drastic action or intervention. The feeble-minded were, as was said, a *menace to society*. This menace was not circumstantial: it stemmed from the nature of the feeble-minded themselves. And it did not elicit the empathy that we typically associate with misfortune or innocence. To paraphrase sterilization survivor Ken Nelson, talking of his time at Alberta's Provincial Training School for Mental Defectives, it was almost as if he and other "mentally defective" children institutionalized there had committed a crime. They were not treated with the empathy and understanding that would typically flow to victims of crimes, but with the anger and retributivism directed at those crime's perpetrators. And as Ken's words intimated, and as the revelations about Alberta's eugenic history were to show, many children who became targets of eugenics did not in fact have the eugenic traits—mental deficiency and psychosis, in the case of Alberta—that were the supposed ground for their treatment. I shall return to Ken's words again in chapter 8, since they provide a key insight into the psycho-social mechanics of eugenics.

That is the eugenic past. But what of the *modern* pursuit of human perfection and its relationship to this kind of subhumanization? Like the old pursuit of human improvement and perfection, the modern pursuit has focused its gaze on the promise that new and emerging technologies hold. At the turn of the preceding century, new sexual sterilization technologies were introduced, such as vasectomy as a surgical procedure, and a range of novel testing technologies—later known commonly as "IQ tests"— claimed to measure a person's intelligence. The turn of the twenty-first century brought with it new concerns about ever more sophisticated, invasive technologies. Of special relevance are two overlapping types of technology. The first is focused on *reproduction*: in vitro fertilization (IVF), prenatal testing in general, and embryo screening and pre-implantation genetic diagnosis (PGD) more particularly. The second concerns *human enhancement*: pharmaceuticals that enhance cognitive capacities, and external devices that replace or extend the limited capacities of the biological body, evoking cyborg futures that call into question, for many, just what we are as a species.

More than twenty-five years ago, the sociologist Troy Duster's *Backdoor to Eugenics* first drew attention to ways in which such technologies facilitate eugenic agendas, even if this was very much eugenics underground. Duster identified two factors that operate to make that back door to eugenics one that remains open in contemporary society: the stratification of society and the power that even the merest whiff of an appeal to biology has to awaken reductionist and deterministic views of human nature within both scientific and popular cultures. Writing in the heyday of the Human Genome Project, Duster very much had the uses and abuses of genetic technologies in focus, especially insofar as they shape "what kind of knowledge we should pursue in determining what kind of children should be born." Making more concrete what he means by the "backdoor to eugenics," and drawing a contrast with the Nazi *Lebensborn* program of positive eugenics that appealed to the enhanced reproductive role of fine Aryans in the future of the race, Duster continues: "To put it metaphorically, when eugenics reincarnates this time, it will not come through the front door, as with Hitler's *Lebensborn* project. Instead, it will come by the backdoor of screens, treatments, and therapies. Some will be admirably health-giving, and that will be the wedge. But sooner or later, each will face the question of when to shut the backdoor to eugenics."

The eugenic potential in these new technologies and our matter-of-fact reliance on them in individual decision making and social policy has been viewed as a kind of eugenic inevitability by the philosopher Philip Kitcher. That potential has also generated debate over acceptable and unacceptable forms of eugenics in the writings of disability theorists such as Tom Shakespeare and bioethicists such as Nicholas Agar. New eugenics, or *newgenics*, is now a live topic in the fragile sciences, one I shall return to and discuss in more detail in chapter 7.

Major disability advocacy organizations, like the Canadian Association for Community Living, know about the back door to eugenics and newgenics from their day-to-day experience, working from within and together with the disability community. And they have had good reason to home in on the modern pursuit of human perfection, and to promote a pro-disability critique of that ongoing pursuit, as a way of questioning status quo views of who is worthy of life. Such views about full human life-worthiness have played a role both in the eugenic past and in the newgenic present, continuing to occupy center stage in fields such as contemporary bioethics. The "Modern Pursuit of Human Perfection" dialogue itself made that much clear to those of us who were more naïve.

Before considering this aspect of contemporary bioethics, I want to spotlight an earlier form of professional advocacy from within medicine that is not as well-known even by historians of medicine as one might expect. It is where euthanasia, eugenics, and disability intersect for the first time in the medical profession.[2]

4.3 Eliminating Defectives in Medicine's Short History

Calls for eliminating people with disabilities were not new with the rise of eugenics in the late nineteenth century. What was new during the short history of eugenics was the growing number of physicians who either expressed support for or even became keen advocates of practices of euthanasia, particularly in the case of infants and children with disabilities. By the time of the First World War, the level of support in the United States for such forms of euthanasia was sufficient for debate over those practices to play out in the public arena. It did so in large part through a popular silent film, *The Black Stork* (1917), featuring the euthanization advocate and practitioner Dr. Harry J. Haiselden.

Haiselden graduated from the Chicago College of Physicians and Surgeons with an MD in 1893, working at Chicago's German-American Hospital as a resident and assistant to the hospital's chief surgeon and president, Christian Fenger, before assuming Fenger's role upon his mentor's death in 1902. Although Haiselden had established a local institution, the Bethesda Industrial Home for Incurables shortly after he gained his MD, he became more critical of institutionalization with his growing interest in eugenics. By the First World War, Haiselden had come to think of the sexual segregation that was part of institutionalization as an ineffective response to the problem of the deficient. In fact, Haiselden appealed to the horrors of institutionalization as part of his explicit justification for euthanizing defective babies and infants.

Haiselden had been euthanizing babies and infants he considered defective from 1915. Moreover, his practice of euthanization had garnered national attention in part through his own efforts to publicize his actions. Haiselden's starring role in a popular silent film was part of that effort. But as the historian Martin Pernick has argued, what was striking about Haiselden's actions at the time was as much his public advocacy as the substance of the views and actions he advocated. By that time, physicians killing "defective infants," or alternatively, deliberately allowing those infants to die through selective nontreatment, was recognized as an increasingly common fact on the ground. As Pernick says, the "extent to which eugenics had come to include the death of the disabled can be seen most clearly in the startling number and diversity of Haiselden's supporters," who came from the general public as well as from within the medical sciences.

In my experience, knowledge of this aspect of the history of medicine itself often comes as a shock to current medical students and doctors themselves, insofar as it comes to their attention at all. Euthanasia ends an individual's life, but it need not be eugenic in nature, even when it selectively targets so-called "defectives." But the attitudes, views, and values that led physicians a century ago to advocate selective euthanasia of babies and infants with apparent disabilities were very much part of a broader eugenic vision that infused at least significant pockets of the medical profession. That vision focused not simply on the life of the individual, but also beyond that life to possible intergenerational effects. And those attitudes, views, and values could occasionally be found vehemently expressed in prominent professional medical journals.

For the most part, medical journals, then as now, are filled with articles, reviews, and discussion pieces whose emotional tenor is flat and objective, as unexciting to read as they likely are unexciting to write. The title "Some Notes on Asexuality; With a Report of Eighteen Cases," from the March 1920 issue of the *Journal of Nervous and Mental Disease*, shares that general tenor. But the article itself quickly departs from the stylistic norms one might expect, revealing in the process much about its author's feelings about sexual sterilization of the feeble-minded. Its author, Dr. Martin Barr, a former president of what is now called the American Association on Intellectual and Developmental Disabilities, begins by informing the reader that sexual sterilization had been practiced across cultures for thousands of years—an appeal to the long past of eugenics—linking this reminder to then contemporary American society by appealing to the alleged fact that "the feeble-minded have so multiplied and increased as to become a distinct race." Barr concludes the notes part of the paper, before moving to his report of the eighteen cases, with another putative fact: the challenge that this new race of defectives posed to American society:

We must face the fact that the very life-blood of the nation is being poisoned by the rapid production of mental and moral defectives, and the only thing that will dam the flood of degeneracy and insure the survival of the fittest, is abrogation of all power to procreate. The shibboleth of the day is "lock up all degenerates once so proven." And this we do. [new paragraph] But sooner or later the brighter ones, whose defects for a time are masked by the benefits received from training, are removed from the protection of sequestration, either by parents or guardians convinced of "cure" so called; or again by the misdirected philanthropy of idle women; or *some* charitable societies, eager to set them at liberty "that they may have their chance." They have it all right! And pressing forward they go out to meet the "Years to Come," and tramp through the black morasses of sexual filth until precipitated into the whirlpool of the stormy "Sea of Life" from which few, if any, ever return: and double prisoners and captives of the vicious, and their own passions, they sink lower and lower, consorting with the ruck and filth, the scum and dregs of mankind. Then these hereditary irresponsibles—degenerates, imbeciles, defective delinquents and epileptics—the very nightmare of the human race, ever with sexual impulses exaggerated, find their "chance" in reproduction. Unconsciously innocent prisoners of a normal race, they are nevertheless its worst enemy.

That this kind of emotionally laden language finds expression not simply in books and journals recognized as vehicles for eugenic propaganda, but also in a respected, mainstream medical journal indicates the extent of the infusion of eugenic thought in early twentieth-century medicine. Like Haiselden's advocacy of eugenically grounded euthanasia from around the same time, Barr's language here may be something that leaves some of us now in our own state of shock. But while that reaction is not without significance, it can become unduly distracting, exciting its own reactionary tirade of dismissal of what is being said. As a counter, I want to return to the two dimensions that anchor my reflections about eugenics as practiced: institutional complicity and the cluster of puzzles about engaged individuality.

As we noted in chapter 2, Barr was the long-serving chief physician at the Pennsylvania Training School for Feeble-Minded Children. Barr had supported an early eugenic sterilization law in 1901 in Pennsylvania, as well as published a book on the treatment and training of the feeble-minded, *Mental Defectives: Their History, Treatment and Training*, in 1904. But Barr's discussion of eugenic sterilization there, while overlapping in content with his 1920 article, shares very little of its vitriol. What sorts of institutional complicity create the ideological and intellectual space for Barr's thoughts, and for their subhumanizing expression in a staid professional journal whose surrounding articles are now merely good bedtime reading for those with a touch of insomnia?

And how is it that Barr comes to manifest such a limit to his engaged individuality when it comes to those who fall under his medical care as a long-serving chief medical officer at a major state institution? That subhumanizing dismissal of institutionalized "degenerates" is of sufficient intensity to tar, in addition, those who show even the slightest sign of that very engaged individuality. What failures of imagination, putative facts, and basic values are operating here? The switch in tone in the passage immediately following the passage already quoted, where Barr goes on to baldly state his own preferences regarding surgical alternatives, only adds to the poignancy of these questions:

In regard to the character of operation: Personally I prefer castration for the male, and oöphorectomy for the female, as insuring security beyond a peradventure; and when performed on the young, desire almost entirely ceases, or is at least held in reasonable abeyance. [New paragraph] If for the sentimental reasons the removal of the organs are objected to, vasectomy or fellectomy [sic] may be substituted.

Here Barr expresses his preference for operations that, unlike vasectomies and tubal ligations, are nonminimally invasive of a person's bodily integrity, referring only to "sentimental reasons" in support of other alternatives.[3]

4.4   Tough Medicine in Postwar Alberta

Haiselden's enthusiasm for the euthanization of defective babies, and the subhumanizing views of mental defectives that Barr expresses, are both a part of the medical profession's entwinement with the eugenic past. In Alberta, the echoes of this tough medicine for "mental defectives" could be heard well into the 1950s and 1960s.

Dr. Leonard J. Le Vann was the long-standing medical superintendent of Alberta's Provincial Training School for Mental Defectives in the post-1945 era, serving in that capacity almost continuously from 1949 until 1974. In that role, Le Vann recommended people at that institution for sterilization, a recommendation that required the approval of Alberta's Eugenics Board. Two troubling aspects of Le Vann's broader practices were revealed during the *Muir v. Alberta* legal case in 1995–1996. They manifest, in other ways, the kind of subhumanizing tendencies that form part of the medical profession's entwinement in the eugenic past.

The first of these is that Le Vann repeatedly used children at the Provincial Training School as subjects in what were essentially experimental psychotropic drug trials during the late 1950s and early 1960s. In particular, Le Vann administered to "mental defectives," including children, who were residents at the Provincial Training School, a range of recently approved medications for people meeting the formal psychiatric diagnosis of psychosis. He did so primarily in order to explore the drugs' effects, although his captive subjects, who were under his medical supervision in the institution, did not meet that diagnosis. These experiments typically were done without parental, guardian, or patient consent, and like the eugenic sterilizations that Le Vann recommended, typically even without the knowledge of the children under his care. (It would take the post-*Muir v. Alberta* legal investigations in the late 1990s to reveal to many of them what had transpired here.) The results of these experimental studies were published in widely read professional journals, such as the *Canadian Medical Association Journal* and the *Canadian Psychiatric Association Journal*. But it appears that these troubling forms of human experimentation drew critical attention as such only thirty-five years later during *Muir v. Alberta*, Leilani Muir's successful lawsuit against the province of Alberta for wrongful confinement and sterilization.

The second is that Le Vann, in concert with Alberta Eugenics Board member Dr. Margaret Thompson, conducted experiments in the early 1960s on the testicular tissue of males with Down syndrome. That tissue had been removed through castration, an operation that itself goes well beyond the less bodily invasive surgery of vasectomy. These castrations resulted from Le Vann's recommendations for sterilization, as the medical superintendent of the Provincial Training School, and Thompson's approval of those recommendations for sterilization, as a member of the Eugenics Board. Strikingly, Thompson was the only geneticist to serve on the Eugenics Board to that point in time, when she was an assistant professor at the University of Alberta. In addition, this cycle of recommendation-approval-experimentation unfolded despite the fact that, as a trained geneticist, at least Thompson recognized that Down syndrome males were very likely already sterile and that the syndrome was not hereditarily transmitted, facts that she later acknowledged during her questioning as a witness in *Muir v. Alberta*. Human experimentation was both psychotropic and surgical in Alberta's eugenic past, and the subhumanized subjects of that experimentation were targets of the province's eugenics program.

The ways in which institutional complicity and the limits of engaged individuality operate together in the subhumanization of those with intellectual disabilities are apparent here, but come out more fully once we attend to Thompson's own views. When Thompson was asked, as a witness called by the government during *Muir v. Alberta*, why she gave such approvals, given that as a geneticist she recognized at the time the sterility of males with Down syndrome, she gave a short response. Thompson indicated simply that sterilization in such a case would "make assurance doubly sure," expressing with this her assumption that the questioning lawyer would surely agree that "nothing much was lost."

But Thompson clearly mistook the state of things in 1995: the courtroom was aghast. Her complacency about the reception of her views was perhaps facilitated by her post-Alberta achievements, which included founding and presidential roles in the Genetics Society of Canada, more than twenty years of service in genetic counseling at the Toronto Hospital for Sick Children, and receipt of the Order of Canada in 1988. Independent eye-witness reports of this part of Thompson's testimony characterize this as one moment during the trial when a recognition rippled through the courtroom that Alberta's eugenics history involved more than misguided good people trying to promote and implement policies of human improvement.[4]

4.5   Rational Ethics: Cognitive Disability in the Eugenic Now

One could explore the workings of institutional complicity and the boundaries of engaged individuality through historical questions about Haiselden and Barr in the United States, or about Le Vann and Thompson in Canada, or about some delineated aspect of the more general history of eugenics. There is much value in completing the detailed, archival, and historically focused research that would partly detail how eugenics does its work in those specific contexts.

But I am more interested at the moment in shifting gears and taking a different tack. I want to show how versions of those very same questions about the workings of institutional complicity and the boundaries of engaged individuality arise in the eugenic *present*, and to consider the implications of this for life, death, and disability *now*. The dimension to The Eugenic Mind that subhumanizes some in order to better human society remains with us. The ongoing advocacy of ideas, practices, and policies in the name of human betterment, like that in the eugenic past, subhumanizes people with at least certain kinds of disabilities. And it does so, in part, I want to suggest, via specific kinds of institutional complicity and various limitations to engaged individuality among advocates of those ideas, practices, and policies. Recognition here of the workings of subhumanization will seed more detailed discussion of aspects of subhumanization in the social mechanics of eugenics in parts II and III.

The growth in reproductive and enhancement technologies in the late twentieth century brought in its train new philosophical specializations—in applied ethics, bioethics, medical and health care ethics, disability ethics. These specializations are dedicated to the moral, social, and political questions raised by these technologies and their implications for individuals and society. While some of the leading figures in these specializations are qualified and work as medical doctors or as professors in medical schools, many have their primary qualification in other fields: for example, in philosophy, in public health, in law, or in sociology. Some of the most prominent of these figures hold and defend views of people with at least certain kinds of disability as having second-class moral status. Although the emotional invective one finds in professional publications by physicians like Martin Barr of a century ago is missing, the positions advocated within contemporary mainstream bioethics share the substance of subhumanizing views of people with disabilities that was prominent among medical professionals in that eugenic past.

Sometimes that subhumanization can be explicit. For example, the dominant view of persons and their identity over time in philosophy views emphasizes the importance of certain kinds of rational capacities. That emphasis also underpins views and principles about moral status and personhood that subhumanize those who lack those rational capacities. Such views and principles have been defended by philosophers such as Peter Singer and Jeff McMahan in part by comparing nonhuman animals to people with severe forms of disability, especially those who fall under the American Association on Intellectual and Developmental Disabilities' category of "profound intellectual retardation." One implication of the comparison—indeed, sometimes the point of the comparison—is that those with profound intellectual disabilities supposedly lack the cognitive capacities required for a living thing to have the moral status *person*, whereas there are some nonhuman animals, such as chimpanzees, that appear to possess such capacities. Those lacking that status do not have the rights that persons have.

Prominent among those rights are the right to life and the right to developmental and bodily integrity. These are protections afforded in a range of international charters and conventions, including the UNICEF Convention on the Rights of the Child (1989, article 6), and the earlier United Nations declarations, the Universal Declaration of Human Rights (1948) and the Declaration on the Rights of Mentally Retarded Persons (1971). For example, in the UNICEF Convention, article 6 states that "1. States Parties recognize that every child has the inherent right to life. 2. States Parties shall ensure to the maximum extent possible the survival and development of the child." On the views of Singer and McMahan, since those with profound intellectual disabilities lack the status *person*, killing them, or allowing them to die, may be morally permissible, while killing or allowing (say) a chimpanzee—a nonhuman animal with the requisite cognitive capacities—to die, is not permissible. The subhumanization of those with profound intellectual disabilities carries with it implications for what is permissible in the treatment of existing people and for how we ought to think about our choices regarding the sorts of people there should be in future generations.

Understanding the historical context in which such views of persons and moral status have arisen, the general and consistent nature of the views articulated, and the intended point of the human-nonhuman animal comparisons, reveals the extensive nature of the engaged individuality of philosophers such as Singer and McMahan. The ratio-centric conception of persons forms part of a long tradition in philosophy, with the most influential philosophical discussions of persons and moral status arising in the early 1970s in debates over abortion focused on reproductive autonomy. Singer and McMahan are committed animal rights advocates and aim to redress what they see as an ascribed status bias, a kind of speciesism, favoring human beings over nonhuman animals, independent of the characteristics that each has. That speciesism, in their view, operates to dismiss concerns about, and the treatment of, nonhuman animals. Their views about the profoundly intellectually retarded are a logical consequence of these broader commitments.

Yet this greater understanding also brings into sharper focus the *limits* to their engaged individuality, and returns us to questions about the kinds of ongoing institutional complicity that maintain such subhumanizing views of some people as dominant views in the field. It is here that standpoint matters, for such views can be articulated, seriously considered, publicly defended, and even come to represent prevalent views only when there is a certain kind of institutional marginalization of those with severe forms of disability, whereby they and those adopting their standpoint occupy at best the periphery of the arena of debate.

It is when that institutional complicity is disrupted that we see emerging challenges to those views on the grounds that they subhumanize those with severe disabilities. Here the standpoint of parents of children with a range of disabilities more generally has played an important role in such challenges. That standpoint has also helped to identify imaginative failure of a certain kind as functioning at both the institutional and individual level. As Eva Feder Kittay has powerfully conveyed, imagine what it is like as a parent—as it has been for Kittay herself for many years—to hear *your child* described by your colleagues and interlocutors as unworthy of life, or of reading articles about such children as having moral status comparable to, or even lesser than, a pig or a dog. There is something subhumanizing in the very idea of rationally justifying or debating *that* for anyone sharing Kittay's standpoint, a point to which I shall return in part III, "Eugenic Voices."

First-person reflections on the unease caused by the double layer of subhumanization involved in confronting this ratio-centric view of persons and moral status sometimes reflect the standpoint of those with the very forms of disability that Singer and McMahan view as depriving an individual of life-worthiness. The disability rights lawyer Harriet McBryde Johnson's *New York Times* op-ed, "Unspeakable Conversations," provides one such reflection. It was written after visiting with Singer at Princeton University at his invitation, a visit that includes giving a guest lecture in Singer's class. The editorial begins with an implicit reference to McBryde's standing as a human being with the kind of developmental disability that would be person-disqualifying on the ratio-centric view of persons that Singer has defended:

> He insists he doesn't want to kill me. He simply thinks it would have been better, all things considered, to have given my parents the option of killing the baby I once was and to let other parents kill similar babies as they come along and thereby avoid the suffering that comes with lives like mine and satisfy the reasonable preferences of parents for a different kind of child. It has nothing to do with me. I should not feel threatened.

There is something subhumanizing about having someone called upon to defend their status as a person, or to rationally demonstrate to someone else where their arguments to the contrary have gone wrong.

If that person were denied full moral status because they were of a certain sex (female), of a certain race (African American), or from a particular ethnic or cultural group (Jewish), both the denial itself and the call to show what is mistaken about the subhumanizing position would readily be recognized as subhumanization. But this is not the case when that denial is based on one's having some kind of cognitive or other disability thought to be serious enough to warrant decisions of life and death. This asymmetry indicates that there remains a deep form of institutional complicity in the subhumanization of people with those kinds of disabilities. It also identifies the kind of limit to the engaged individuality of those advocating views as being clearly unacceptable forms of discrimination: sexism, racism, and ethnocentrism.

With much of the focus here being on parental choice, children, and disability, the subhumanizing discussions readily encompass an explicit intergenerational dimension. Consider one principle that has been articulated by the Oxford-based bioethicist Julian Savulescu, a principle that I shall return to discuss in more detail in chapter 7. Savulescu advocates what he calls the *principle of procreative beneficence*: "couples (or single reproducers) should select the child, of the possible children they could have, who is expected to have the best life, or at least as good a life as the others, based on the relevant, available information."

Savulescu articulates the principle of procreative beneficence in the context of parenting decisions using in vitro fertilization and pre-implantation genetic diagnosis. But as a general principle, its discussion and defense ranges beyond parental choices concerning embryos. Although Savulescu's initial focus is on the implications that the principle has for nondisease traits, such as intelligence, he later considers more explicitly its implications for disability.

Like the discussions of Singer and McMahan, much of what Savulescu says is thoughtful, honest, provocative, innovative, and respectful. This includes postulating an asymmetry between abilities (such as being able to hear) and disabilities (such as being deaf) with respect to what Savulescu calls "opportunity restriction," and sketching an argument linking the nature of autonomy to the selection of children with the best lives. Like those discussions, Savulescu's principle is premised on the relatively uncontroversial view that there are at least some better and worse states for a person to be in, together with the more questionable view that states of disability are clearly worse states in many contexts. But for all that, the substance of the view that Savulescu defends, like that of Singer and McMahan, subhumanizes people with at least some kinds of disability in ways that are continuous with the eugenic past.

Consider one of Savulescu's concluding statements of his view: "we do have good reason to select the best children, those with the least disabilities. We should no longer leave reproduction thoughtlessly to chance." That conclusion could have found a comfortable place among early pro-eugenic writings, were "disabilities" simply replaced with "defectiveness."[5]

4.6 Subhumanization and Standpoint's Complexities: Ashley X

Thus far in this chapter I have focused on what I have called the subhumanization of those with intellectual and cognitive disabilities, particularly as it operates through institutional complicity and limits to the engaged individuality of those in positions of epistemic authority and power. I have also pointed to disruptions of that complicity posed by first-person perspectives on what are typically regarded as "severe disabilities," such as those of Harriet McBryde Johnson, and by those who occupy a similar standpoint in their role as caring parents of those so regarded, such as Eva Feder Kittay.

Yet the role of both first-person and parental perspectives within a standpoint eugenics are more complicated than the posited contrast that I have drawn suggests. Many parents in the eugenic past institutionalized their children and consented to their sterilization because they believed that this was in the best interest of those children, as well as their family or broader society. As indicated by the recent Senate inquiry in Australia into the ongoing sexual sterilization of girls and women with intellectual disabilities mentioned in chapter 1, that view is still present today, even if it is not as prevalent as it was. More generally, what might be regarded by some as the subhumanizing treatment of the disabled, particularly the intellectually disabled, continues to be adopted by, even actively advocated by, parents, guardians, and others who believe themselves to be acting in the best interests of their children or wards.

One case that has garnered much attention here is that of "Ashley X." Ashley has profound developmental disabilities, including cognitive disabilities, though just how severe these latter are remains a matter of some dispute; Ashley is now in her late teenage years, living in the Seattle area. Ashley is nonambulatory and requires high levels of care provision. Ashley came to international attention in 2006 not because of anything intrinsic to her condition, but because in 2004, when she was six, her parents and the Seattle Children's Hospital decided to modify her body in ways that would stem her normal pattern of growth, and more particularly, delay and minimize Ashley's sexual development. The resulting "growth attenuation treatment" included subjecting Ashley to large doses of estrogen, as well as surgical interventions on her body, such as performing a hysterectomy and a double mastectomy, later called "breast bud removal" by her parents and the doctors involved in the treatment.

Discussion of Ashley's case was widespread in the bioethical and medical communities following a journal article by her doctors and a blog run by Ashley's parents, both of which reported her case and offered justifications for her medical treatment. Through those reports, blogs, and ensuing discussion, Ashley came to be known as the *pillow angel*, a term used initially by Ashley's parents, since Ashley would remain wherever she was placed, often on a pillow. One primary justification offered for these interventions on Ashley's bodily integrity was that growth attenuation treatment, by keeping Ashley small, would make it easier to keep her integrated into family life, both at home and on family trips together, and thus increase the quality of her life. The overall cluster of medical interventions on Ashley's body were justified in terms of Ashley's own best interests, including those that keep her integrated into her existing family environment and so beyond the reach of institutionalization. By confining the growth of her body through biochemical means, it seems, Ashley would be less likely to face the confinement of institutionalization.

High estrogen dosing to affect growth and development has formed part of growth attenuation treatments practiced on children, especially girls, who were unusually tall, since 1956, in order to bring their childhood and adult heights closer to average heights. However, Ashley was the first-known person with cognitive disabilities to be subject to this treatment in virtue of the supposed severity of those disabilities. From the outset, part of the justification for Ashley's specific treatment was the advocacy of growth attenuation as a general treatment for children with severe cognitive disabilities whose normal bodily growth would pose the same kinds of issues for families as they did in Ashley's case. Indeed, in the original paper reporting Ashley's treatment, the authors advocate that for children with "profound developmental disabilities ... after proper screening and informed consent, growth-attenuation therapy should be a therapeutic option available to these children should their parents request it." In fact, what is now known as the "Ashley treatment" had been used on more than a dozen children by 2012, as reports in the UK newspaper *The Guardian* revealed.

Questions about what are in the best interests of a child are not always easy to answer, and parents of children with disabilities are often placed in situations that are intrinsically complicated here. Parental views in the eugenic past are readily seen as having been influenced, even shaped, by a medical world that was sometimes explicitly pro-eugenic, as we saw in the cases of Harry Haiselden and Leonard J. Le Vann, but more often implicitly so, exemplifying institutional complicity in eugenics. The beliefs of many parents who held that it was in the best interests of their child to be institutionally segregated and sterilized were influenced, in part, by the advocacy of medical and other experts and authorities to whom they deferred.

Yet this kind of influence is not always operative, and there are many reasons why parents can differ in their views of whether certain interventions on the body of their disabled child are in that child's best interest or are likely to result in an improvement in their quality of life. This implies that a standpoint eugenics incorporating the perspectives of well-meaning parents is unlikely to have any more unity to the views it expresses than there is with respect to views about the constituent issues that they must address more generally. Nonetheless, there are features of Ashley's situation that suggest that her treatment was subhumanizing, regardless of parental intent. If that is correct, then advocating for that treatment stands in tension with any disability-sensitive standpoint.

First, while the doctors involved in the treatment endorse the inclusive belief "that all children, regardless of the presence of a disability, belong in families" and Ashley's parents, as well as other parents of severely disabled children who have undergone growth attenuation treatment, have expressed the view that they are doing for their child what *any* loving parent would do for their child, this apparent support for equal treatment seems inconsistent, limited, and even opportunistic at times. For example, while the safety of treatment untested on children with severe disabilities is justified by the doctors by an appeal to its use on non-severely disabled children, the parents have objected to the requirement of judicial review of any sterilization operation as imposing too heavy a burden on parents of children who are severely disabled. To be clear, the procedures Ashley was subject to were illegal under Washington law, violated article 6 of the United Nations Convention on the Rights of the Child (specifically the right to "survival and development"), and bypassed the usual ethical protocols requisite at the servicing hospital. It seems likely that procedures subject to these kinds of legal and ethical irregularities that were performed on a child without severe cognitive disabilities would almost certainly have been met with criminal charges being laid against either the parents or the hospital (or both). As such, far from being justifiable by an appeal to a child's right to family life or as an expression of parental love, this likely outcome calls into question those very appeals.

Second, and more generally, judgments about whether specific treatments, especially those involving bodily interventions with lifelong effects, are justified in the interests of a severely disabled child should meet something like a Rawlsian *veil of ignorance* requirement, a graphic way to express a basic parity principle. Any bodily intervention that is justifiable should be justifiable *whether that child is severely disabled or not*, meaning that decisions about bodily intervention on a child should be conducted behind a veil of disability ignorance. As Eva Feder Kittay has expressed this idea with respect to growth attenuation specifically: "If growth attenuation should not be done on children without these impairments, then it should not be done on any children. To do otherwise amounts to discrimination."

Even simply taking departures from this veil of ignorance constraint to place a high burden of proof on those appealing to "severe disability" as the grounds for the departure raises questions as to whether that burden of proof was met in the Ashley case, despite the extensive accounts provided by her parents. In a world in which selecting children with a minimization of their disabilities is seen as a parental obligation, it is easy for parents to overlook ways in which their advocacy for their children repeats, in a new guise, much of the eugenic past.[6]

4.7   Life-Worthiness and Human Variation

This chapter began by reporting a public dialogue on human perfection and life-worthiness that was sponsored in part by the Canadian Association for Community Living (CACL) in response to perceptions of disability, parenting, and the lives that are worth creating. As I sat down to revise the chapter, that same dialogue was being continued in debate over disability, illness, and the termination of lives deemed no longer worth living. That debate has been fueled by Canada's recent physician-assisted suicide legislation, Bill C-14, passed in Parliament between April and June 2016 following the Canadian Supreme Court decision, *Carter v. Canada* (AG) in February 2015 that found that the law which had, until that point, banned assisted suicide was unconstitutional and in violation of the Canadian Charter of Rights and Freedoms. Part of the reaction to the court decision and legislation within the disability community represented by the CACL has been to counter dominant narratives about disability and the value of life with the video stories developed by community activists in Canada as part of Project Value. These short narratives aim to shift the marginalized perspective of people with disabilities, and people parenting with disabilities, to center stage. The epistemic change that this kind of standpoint eugenics will have on public perception and the uptake of Bill C-14 remains unclear.

I have suggested that the entwinement of the medical profession in the eugenic past of the early twentieth century raises questions about both institutional complicity and limitations to engaged individuality. That entwinement is also manifest in Alberta's postwar eugenic history, and I think the same questions about institutional complicity and engaged individuality arise in that context. Furthermore, such questions also can be and ought to be raised with respect to the ratio-centric ethics that dominates much of contemporary bioethics and that partially structures contemporary discussions of newgenics and disability.

By adopting at least a standpoint-ish eugenics, *The Eugenic Mind Project* has so far highlighted the historical continuities between an acknowledged subhumanizing past and views that are very much a part of the current academic landscape. In so doing, it casts our ongoing discussions and debates in a new light. I have not tried here to argue thoroughly for or against any of these views, or sought to answer in-depth questions about the "how" of institutional complicity and engaged individuality in particular cases.

Our discussion in chapter 5 will take us closer to such arguments and answers, which will be developed more fully throughout the chapters in part II, "Eugenic Variations." Here I turn to the issue of how human subnormalcy and disability themselves have been conceptualized, particularly via their medicalization, and the historical and conceptual grounds for our ideas about human variation.[7]

II Eugenic Variations: The Persistence of Eugenics

5   Where Do Ideas of Human Variation Come From?

5.1   Standpoint, Prosociality, and Human Variation

As emerged in part I, "Eugenic Activities," the kind of standpoint eugenics being developed here takes the continuities among ideas, practices, and policies in the eugenic past and those in the newgenic present to be substantial. This perspective reflects the standpoint of at least two types of eugenics survivor: survivors of practices, such as institutionalization and sterilization, from the explicit eugenic era, and survivors of the legacy of that era, such as those living (and particularly parenting) with disability now. In part II, "Eugenic Variations," I turn to consider the social mechanics of eugenics. Here our preoccupation will be with the persistence of eugenics over time.

In chapter 1 I also located eugenics—perhaps, for some, more surprisingly—against a backdrop of human prosociality, our tendency as a species to be cooperative, helpful, and altruistic toward one another. There I also suggested that our extreme vulnerability in infancy and childhood, and the variety of social dependencies that are a function of that vulnerability, provided reasons to adopt such a view of human nature. The engaged individuality that each of us displays is, in part, prosocial and derives from our origins as the kind of creature whose existence presupposes that prosociality. That too I take to be a lesson of standpoint-ish eugenics itself, something that the survivor perspective makes perspicuous. We can return to this point about prosociality and vulnerability in reflecting on human variation, the focus of this chapter.

Each of us is born as a relatively small, highly dependent and vulnerable creature. Without a prosocial milieu, we would not survive more than a few days. In the normal course of events following our birth, our mere presence elicits good will and active care from those on whom our continued existence depends. We are nourished, we are protected, we are integrated into one or another kinship and community structure. Were we not fed and sheltered, nurtured and welcomed, our dependence and vulnerability would instead be markers of a life brutish and short. From our initial prosocial reception, our individual life trajectories are contingent on our genetic and biochemical constitution, the character of our immediate physical environment, the affective care that we receive from those in closest physical proximity to us, the opportunities for education and learning provided in our society, and the spaces created and limitations set by the social categories that we fall under in our specific cultural circumstances. From this vulnerable origin, human individuality, in all its variety, flourishes, thanks to our prosocial nature.

Like much variation in the natural world, human variation is truly something to marvel at. The dimensions to this variation are many: developmental, social, cultural, genetic, psychological. Under the heading "human diversity," human variation is often thought of primarily as *cultural variation*. As such, it is the focus of much work in the social sciences, especially in human geography and in all four fields of anthropology, but especially in cultural anthropology. Here the extent of human diversity is documented, speculated about, and celebrated. An understanding of human variation is central to many areas within the fragile sciences.

Yet despite its positive, crucial role in individual flourishing, cultural variation or diversity has not always been met with the warm embrace of prosocial inclusion. As we saw in chapter 1, eugenics as the science of human improvement both reflected a prosocial tendency and expressed the limitations to that tendency with respect to whose lives were to be improved. The boundaries to our engaged individuality with respect to race, ethnicity, and culture is something known by anyone even passingly familiar with the history of racism, colonialism, and ethnocentrism. In chapter 2 we identified some of the ways in which eugenics has functioned with respect to race and ethnicity, intersecting with that history. But the limits to our prosociality exist with respect to human variation more generally, as much with respect to the genetic, developmental, behavioral, and cognitive variation in our species as with racial, ethnic, and cultural variation.

As we saw reflected in the prominence of feeble-mindedness and insanity as eugenic traits in chapter 3, cognitive variation became particularly important early on within the eugenics movement, particularly that which putatively signaled mental deficiency or illness. Against the backdrop of our prosocial nature, eugenics comes to be understood as a manifestation of the limits of our engaged individuality particularly with respect to *cognitive* variation. With its aim of human improvement, the eugenics movement created particular ways in which both institutions and the individuals through which the practice of eugenics operated became complicit in limiting engaged individuality. The construction of The Eugenic Mind became its own eugenic mind project.

In chapter 4, I identified less familiar ways in which eugenic thinking has been extended to structure contemporary theorizing within the fragile sciences, focusing on advocacy and attitudes within bioethics and medicine in particular. The subhumanization of mental defectives within medicine and the health sciences represents one kind of response to the cognitive dimension to human variation in the eugenic past. The subhumanization of those with profound mental retardation within bioethics represents a contemporary response to cognitive variation that signals—from the standpoint of people with and thought to have those and cognate disabilities— the ongoing eugenic present.

As the beginning of a more sustained discussion of the persistence of eugenics, in this chapter and the next I turn more squarely to human variation and the desire to regulate it, to evaluate it, to taxonomize it. Here I am interested not so much in human variation itself but in our attitudes and reactions to that variation, especially those that are less celebratory in nature. Those attitudes and reactions represent a puzzle about human variation. Understanding the puzzle and responses to it are informed by, and in turn inform, the subhumanization we have discussed in chapter 4. I call the puzzle *the puzzle of marked variation* and it will be at the center of this chapter and the next.[1]

5.2   The Puzzle of Marked Variation

Some differences between people are not noticed or registered at all in our everyday interactions with one another. Examples include variation with respect to the third letter of a person's surname, the distance between a person's left shoulder and right knee when walking, and the precise number of hairs on a person's head. Other differences are more readily registered, but typically they are taken to have little robust significance—for example, the length of one's surname, the rapidity of one's ambulatory gait, and how much head hair a person possesses. And further differences are both readily registered and taken to be significant enough to call to mind the presence of a different kind or sort of person. The puzzle of marked variation concerns this third kind of human variation.

This latter kind includes differences of ethnicity, race, sex and gender, and sexual orientation. Also included are the various categories associated with disability, such as impairment and handicap, that I will consider together as *categories of disablement,* as well as those grouped under the pathologizing medical categories *disease, syndrome,* and *condition.* Although the puzzle of marked variation could be expressed quite generally about any sorting of human beings—by ethnicity, class, country of origin, race, gender—I want to focus here on the form of this puzzle that seems to me most closely related to the history of eugenics, and most relevant for thinking about the relationship between that history and ongoing forms of exclusion, discrimination, and advocacy that cluster under the heading of newgenics. Thus, I will be focused on *negatively* marked variation, and on categories of disablement and the previously cited pathologizing sortings of people. And extending our initial probing of eugenics in part I, I shall concentrate more specifically on the medicalization of forms of human variation that tend to be viewed as falling below the normal: instances of *subnormalcy.*

The puzzle of marked variation is this. Human variation is ubiquitous. Some of this variation is registered by us, individually and culturally, as difference. And some of these registered differences are marked in ways that imply that they matter more than do other differences. In fact, they matter sufficiently that the categories we use to register and communicate about these differences operate as categories for different sorts of people. A group of such marked differences especially pertinent in the eugenic past cluster under categories of *disablement* and *medicalized pathology.* These categories pick out individuals who are regarded as being, in some medically significant way, subnormal. What is it that provides the basis for our registration of disablement and medicalized pathology as forms of marked, subnormal variation?

I have stated the puzzle of marked variation to facilitate discussion of the relationship between any solution to this specific version of the puzzle and its generic variant. My own solution to the puzzle, elucidating that relationship, will be introduced in chapter 6. But moving from the generic to the particular articulations of the puzzle that I shall focus on in this chapter, we can express the puzzle in one final way by drawing on the standard philosophical distinction between ontology, the study of what there is, and epistemology, the study of what we know. Ontologically, human variation is ubiquitous, while negatively marked difference is not. Human variation saturates our world in its physical, biological, cognitive, social, and cultural dimensions. Negatively marked difference, by contrast, is arguably not a feature of that world itself at all. Epistemologically, however, it is marked difference that is everywhere for us, with its remainder, mere variation, noticed only secondarily, if at all. Marked variation is often how we identify ourselves and others, and it structures the prosocial lives that we each lead. Given the ubiquity of human variation in the world, what is it that creates the epistemic possibility of marked variation? In the case of disablement and medicalized pathology, what makes it epistemically possible, in a world full of many distinctive forms of human variation, for us to mark some of that variation as subnormal?[2]

5.3　Four Initial Desiderata

We might thus think of this as a puzzle about the epistemology of disablement and medicalized pathology. One of the aspects of this puzzle that seems especially striking concerns the phenomenology of marked variation, particularly that of disablement. Here I mean what it is like to experience the world through the registration and detection of marked variation, particularly disability. The marked variation of disablement has at least two dimensions of *felt immediacy* to it that registration of mere difference or bodily variation does not. One concerns the input side of this registration of marked variation, the other concerns the output side.

First, the detection of, or sensitivity to, marked variation has the same kind of feel to it that visual perceptual identification has in being typically unmediated or direct. When we think that we see some particular thing— for example, an apple before our eyes—we simply *see what it is*, rather than consciously use our background knowledge and reasoning skills to figure out what it is, however much perception depends on prior categorization and conceptualization. Detecting marked variation feels like that: it is phenomenologically direct and unmediated. It is something that strikes us as direct and unmediated.

Second, marked variation has an emotional grippiness to it, a hold on how we respond in the first instance. In this respect, it is much like the emotional experience of fear and disgust and the bodily grippiness they each have on us, in the first instance. Marked variation elicits a kind of emotional reflex inside us. This internal reflex might not lead to any particular behavior on any specific occasion, and it is a reaction that we can learn to dampen or control. But it remains nonetheless phenomenologically salient. In short, we don't simply register the marked variation of disablement transparently, but also respond to it emotionally with the same kind of directness and immediacy.

Together with our preceding discussion, this suggests four desiderata that any adequate response to the puzzle of marked variation should meet. Our response to the puzzle of marked variation should:

1. Recognize the puzzle of marked variation as an ongoing concern, rather than one arising only for an explicitly eugenic past;
2. Ensure that the account of marked variation it provides is compatible with our general prosocial nature as a species, and explain how the two are to be integrated;
3. Explain how various manifestations of the puzzle of marked variation— those arising with respect to race, ethnicity, and disability, for example— are related to each other;
4. Account for the felt immediacy to the phenomenology of marked variation, addressing both dimensions to this felt immediacy, the one concerning perceptual input, the other concerning emotional output.[3]

5.4   Variation, Subnormalcy, and Categories of Disablement

Having noted already that my interest here is in negatively marked variation, I have gone on to focus that interest on categories of disablement and medical pathologization, and to express this more specific interest in terms of the concept of subnormalcy. In this section I want to say more about that concept and its relationship to categories of disablement.

Subnormalcy can be found for both discrete and continuous properties and, at least in the latter case, conventionally determined levels segment a continuous reality into discrete categories. This segmentation is often codified in the law (e.g., *legally blind*), or in criteria for medical categories (e.g., *profound mental retardation*). Subnormalcy is signaled, often powerfully, by appearance, by behavior, by independence in everyday life, and by the everyday and medical categories and labels used in describing a person.

Subnormalcy is pervasive in these signals, even if it is not always explicitly named. To take one quick example, the concept of subnormalcy plays a role in a common way of distinguishing treatments, therapies, or cures—which aim to return someone from the subnormal to the normal—and enhancements—which aim to move someone beyond the normal. Treatments, therapies, and cures, so it is said, "fix a problem" or "make someone well"; enhancements make someone "better than well," to use the title of Carl Elliott's better than well-known book. Wellness, or health, is the normal condition either to which one is restored—via treatment, therapy, or cure—or beyond which one is taken—via enhancement.

To take a more detailed example, consider the widely used World Health Organization's International Classification of Impairments, Disabilities and Handicaps (ICIDH). The ICIDH, which derives from medical models of disease and has come to play a central role in the thinking of major health and disability organizations, attempts to distinguish physical, functional, and social aspects to disablement through the following characterizations of impairment, disability, and handicap:

Impairment: "any loss or abnormality of physiological, psychological or anatomical structure or function"

Disability: "any restriction or lack (resulting from an impairment) of an ability to perform an activity in the manner or range considered normal for a human being"

Handicap: "a disadvantage for a given individual, resulting from an impairment or a disability, that limits or prevents the fulfilment of a role that is normal (depending on age, sex, and social and cultural factors) for that individual"

As the ICIDH goes on to make clear, handicap is "a social phenomenon, representing the social and environmental consequences for the individual stemming from the presence of impairments and disabilities." We can briefly illustrate this trichotomy in the case of vision.

Consider a person who has damage to her cornea that makes it much harder for her to rely on vision as a mode by which she navigates the world, including moving, identifying objects, and reading. In this case, damage to the tissue in the cornea is an *impairment*; significantly reduced visual functioning is a *disability*; and more limited access to books and other written material is a *handicap*.

Implicit in this trichotomy is a kind of hierarchical relationship. The handicap results from the disability, given a social context in which reading material is typically and predominantly available only in a visually identifiable form. It is in this sense that the disadvantages of the impairment or disability are "created by" their social reception, and thus that handicaps for those with certain disabilities could be addressed primarily by changing this social reception. Likewise, the disability results from the impairment, given a bodily context that functions on certain kinds of inputs to, and outputs from, the cornea. We can, of course, modify our bodies so as to change how they function, including in compensatory ways that are adaptive to malfunction, failure, and damage of various kinds. Yet there is a clear sense in which disabilities such as reduced visual functioning are not as readily modifiable as handicaps are by changing social reception and social context.

We might also conceptualize this trichotomy in terms of the *subject* of the corresponding predication, the thing to which an impairment, disability, or handicap is attributed. Doing so can help to further elucidate the relationship among these three categories of disablement. A *body part*, such as tissue in the cornea, is impaired, but it is a *person* who has the resulting disability, contingent on how that impaired body part functions in the bodily systems it plays a role in. And that person will also face a handicap, contingent on the social context in which he or she is located. A person with a disability need not face a handicap if that person's social context meliorates the negative effects of that disability, just as an impaired body part need not contribute to a disability for a person if that person's body adjusts in compensating ways. Disability is thus not intrinsically handicapping any more than impairment is intrinsically disabling.

The first point to draw attention to is the explicit reference to *normalcy*, and departures from it, in all three components to disablement. The second is that, as the talk of loss, abnormality, restriction, lack, and disadvantage implies, the deviations from normalcy that each involves are not of neutral valence but are negatively marked. Hence, each involves, at least implicitly, an appeal to a concept of *subnormalcy*. Third, the augmentation of this framework that adds *susceptibility* (or some variant, e.g., at risk for, disposed to) is subject to both of these points, and so also relies on a notion of subnormalcy. A fetus that encounters a chemical substance in utero (e.g., thalidomide) that alters its normal development, or that contains additional biochemical material in its cells that results from extra-normal chromosomal duplication (e.g., trisomy 21), might be said to be susceptible to, or at significant risk of having, an impairment or disability. This is so even if, at the point at which this judgment is made, that fetus does not have impaired parts or a disability, as defined by the ICIDH.

Back to the question in this chapter's title: Where do ideas of human variation come from? I have already said that I am interested here in the difference between mere variation and marked variation, in versions of this question pertaining to health and medicine, and to disability in particular, and in deployments of ideas of variation that signal that difference. Thus one re-expression of this question has been: What is it that provides the basis for our registration of disability and medicalized pathology as forms of marked variation? Since the puzzle is one in part about how it is that marked variation jumps out at us from a sea of variation, a second re-expression of the puzzle is this: given the ubiquity of human variation in the world, what is it that creates the epistemic possibility of marked variation?

In this section I have argued that the concept of subnormalcy is pervasive in at least the views of disability that are most entrenched in medicine and the health sciences. Since disablement is a major category that marks variation-with-difference in medicine and the health sciences— just as disease and other types of medical pathologization do—we can re-express the question I am asking in a third way: Where do the ideas of normalcy and subnormalcy that mark disablement and medical pathologization come from?

In section 5.5, I want to provide an overview of the answer to this question and more generally to the puzzle of marked variation that has become ascendant in disability studies through the work of influential scholars who acknowledge an explicit debt to Michel Foucault, such as the disability studies scholars Lennard Davis and Shelley Tremain, and the sociologist Nikolas Rose. Broadly speaking, this answer stems from the recognition or insight that categories of disablement are *socially constructed* categories, and aims to understand those categories in terms of a genealogical history that appeals to the idea that the phenomenon of disability was socially constructed through late nineteenth-century eugenics.[4]

5.5   Biopolitics and the History of Eugenics

The short form of this constructivist answer appeals to the central role that the notion of normalcy played in shaping up how disability came to be conceptualized. Using the terminology that Foucault introduced and that is taken up in various ways by Davis, Tremain, and Rose, on this view the ideas of normalcy, and so subnormalcy, that mark disablement derive from a distinctive kind of *biopower* or *biopolitics,* one that arose relatively recently and has provided a way of *disciplining*—regimenting, making docile, regulatively improving—certain kinds of bodies. In contrast to the notion of power familiar from liberal political theory, according to which power is imposed by a centralized source (e.g., a state) on individuals, biopower operates in a more diffuse, self-regulatory manner, depending very much on the uptake of key ideas, practices, and institutions from those subjected to that power. Biopower is, in short, a matter of what Foucault calls *government,* a form of social control that strategically shapes the range of actions of individuals and groups of individuals without instigating a confrontation between those on different ends of the power relation. The introduction of the ideas of normalcy and subnormalcy as marking disablement formed a part of the way in which biopolitics socially constructed disability. It did so by operating through certain social institutions, practices, norms, and values that arose at a specific moment in time. Proponents of this general view differ in their more precise accounts of how such biopolitics and disciplining have functioned.

For example, Leonard Davis argues that "the categories 'disabled,' 'handicapped,' 'impaired' are products of a society invested in denying the variability of the body" (xv), and locates his view in a kind of Marxist understanding of class and industrialization that identifies the middle of the nineteenth century as the crucial time of this biopolitical nexus. Shelley Tremain endorses a form of nominalism about ontological commitment in order to argue, against proponents of both the so-called biomedical and social models of disablement, that attributions of *impairment,* and not just disability (or handicap), involve the regimenting, making docile, and improving of the subject of disablement. Tremain anchors her critique in an analogous critique of the common invocation of the distinction between sex (as natural) versus gender (as socially constructed): sex, she argues, is socially constructed as much as is gender. Likewise with impairment, Tremain claims, with that social construction being mediated by "technologies of normalization" that objectivized individuals through nineteenth-century clinical practices. Nikolas Rose, by contrast and as we saw in chapter 2, concentrates on the history of psychology in Great Britain and its relationship to eugenics. Here Rose identifies the *psychological* normalization of human variation in mental abilities as the critical way in which individuals came to be disciplined by a form of what he calls *psycho-eugenics* that operated primarily in the early twentieth century.

Despite the different emphases—on the body, on medicalization, and on psychological normalization, respectively—there is a common idea at play here. That idea is that normalcy itself is a kind of social construction, rather than something to be found in the world. Its emergence is tied to specific regimes of power, or what Foucauldians typically call power/knowledge, that discipline bodies and minds in particular ways. We should thus think about normalcy (and so subnormalcy) not as a given, but as the result of an activity, that of *normalizing*, that arises within an existing network of power. Normalizing operates to discipline bodies and govern minds in certain ways, and therefore plays a significant role in constituting those subjected to this disciplining, the disabled. Since Davis offers the clearest, most sustained, and influential development of this view of normalcy, variation, and disablement, particularly in its relation to eugenics, I will focus on his views in what follows.

Both Davis's *Enforcing Normalcy: Disability, Deafness, and the Body* and his edited collection, *The Disability Studies Reader* have been very influential in consolidating disability studies as an academic field within the humanities and social sciences. In chapter 2 of *Enforcing Normalcy*, a chapter reprinted in the second edition of the anthology as "Constructing Normalcy: The Bell Curve, the Novel, and the Invention of the Disabled Body," and in its most recent edition as "Disability, Normality, and Power," Davis supports his earlier claim that disability "is really a socially driven relation to the body that became relatively organized in the eighteenth and nineteenth centuries" (3) by arguing against what he calls "a common assumption" about the concept of normalcy:

A common assumption would be that some concept of the norm must have always existed. After all, people seem to have an inherent desire to compare themselves to others. But the idea of a norm is less a condition of human nature than it is a feature of a certain kind of society. ... the social process of disabling arrived with industrialization and with the set of practices and discourses that are linked to late eighteenth- and nineteenth-century notions of nationality, race, gender, criminality, sexual orientation, and so on. (24)

After discussing these practices and discourses, intertwined as they are with the early part of the short history of eugenics, Davis concludes that the "very term that permeates our contemporary life—the normal—is a configuration that arises in the particular historical moment. It is part of a notion of progress, of industrialization, and of ideological consolidation of the power of the bourgeoisie" (49).

Davis's claim here parallels Foucault's own celebrated constructivist claims about social phenomena such as madness and homosexuality. Like the latter of these claims, Davis's claim about normalcy is that it is an idea that arose in the nineteenth century in the industrial West. Davis pegs this more specifically as located between 1840 and 1860, a sort of necessary precursor to Galton's eugenic moment. Although there is a temptation—one that Davis himself sometimes flirts with—to qualify what Davis is saying here by replacing "normal" with "normal *body*," much of Davis's discussion is cast not only in terms of a general notion of normalcy, but even more generally in terms of *norms*.

So contrary to what one might initially think, claims Davis, normalcy is a relatively recent idea, one made possible by certain large-scale societal changes. According to Davis, prior to the middle of the nineteenth century, views of human bodies were governed by *ideals*, typified by ancient Greek and Roman statues that depict perfect bodies, bodies that we can only approximate but never instantiate. These ideals or excellences—in shape, size, proportion, and functioning—might be something to aspire to or admire, but because they are unattainable there is "in such societies no demand that populations have bodies that conform to the ideal" (25). Much the same could be said of the idea of excellences of the mind or soul (*psyche*) in ancient Greek and Roman societies, of the virtues of character and behavior. Since "the concept of the norm or average enters European culture ... only in the nineteenth century" (25)," and came to discipline bodies and minds alike in particular ways, one needs to understand the *genealogy of normalcy* to grasp how disablement has come to occupy the place it does in both ordinary discourse and in medical discourse.

Davis follows Donald Mackenzie's *Statistics in Britain* in viewing the development of statistics, eugenics, and disability as entwined in the genealogy of normalcy. In particular, the development of the Bell curve, standard deviation, and other statistical concepts, and the role that they played in regulating populations of people, constructed certain kinds of conceptual and technological regimentation of individuals with bodies that varied. These allowed certain kinds of representation of normalcy that could then be used to discipline individuals and their bodies through measurement, classification, modification, segregation, improvement, and restriction. They led to the invention of the disabled body.

Davis's view provides a specific social constructivist view of eugenics and disability, as well as a corresponding answer to the eponymous question of this chapter. But the general genealogical perspective on normalcy as a nineteenth-century creation is shared not only by Tremain and Rose but also much more generally.[5]

5.6 Evaluating the Appeal to Biopolitics

How plausible is this answer to the general question "Where do ideas of human variation come from?" A more precise yet cumbersome specification of this question is "Where do the ideas of normalcy and subnormalcy that mark disablement and medical pathologization come from?" I think that while the appeal to biopolitics at the heart of the social constructivist answer tells us something important about eugenics, it is more limited as a response to these questions and the puzzle of marked variation.

As this re-expression of the puzzle suggests, the social constructivist answer is right to identify the concepts of normalcy and subnormalcy as central to understanding the puzzle. What is also true is that from roughly 1860 to 1920 there was a confluence of ideas, techniques, institutions, practices, laws, organizations, and figures of influence that resulted in a particular social construction of normalcy being used in the eugenic configuration of disablement. And it remains a profitable approach to disablement *now* to adopt the general Foucauldian stance that one finds in authors such as Davis, Tremain, and Rose. Tremain's critique of both sex and impairment as natural categories standing apart from the biopolitics that has socially constructed the categories, respectively, of gender and disability has been rightly influential. And Rose has produced some of the most thought-provoking and interesting recent work in the general area, such as his *The Politics of Life Itself*, focused on biotechnology, transhumanism, and neuro-ethics, and his more recent *Neuro* critically examining the rise of the neuro-sciences and their place in the understanding of human nature.

Yet the central claim here—that normalcy and even normativity itself originates in some kind of nineteenth-century biopolitical nexus—is less plausible as an answer to the earlier questions about human variation and subnormalcy. As a start on seeing why, consider the four desiderata that we introduced in section 5.3. Recall an adequate response to the puzzle should:

1. Recognize the puzzle of marked variation as an ongoing concern, rather than one arising only for an explicitly eugenic past;
2. Ensure that the account of marked variation it provides is compatible with our general prosocial nature as a species, and explain how the two are to be integrated;
3. Explain how various manifestations of the puzzle of marked variation—those arising with respect to race, ethnicity, and disability, for example—are related to each other;
4. Account for the felt immediacy to the phenomenology of marked variation, addressing both dimensions to this felt immediacy, the one concerning perceptual input, the other concerning emotional output.

The first of these desiderata highlights a strength of the appeal to biopolitics, where the legacy of eugenics can be readily understood to be a far-reaching consequence of the effects of a nineteenth-century biological nexus that socially constructed disability as marked variation in terms of the notion of normalcy. Since that notion remains firmly embedded in our current conception of disability, as section 5.4 suggests, the appeal to biopolitics has identified something important connecting the eugenic past to the newgenic present.

However, this response to the puzzle of marked variation does little by way of satisfying any of the other desiderata here. Although there is nothing strictly incompatible between the appeal to biopolitics and the second desideratum, the fact that that appeal floats free of *any* commitment about idea of what it is to be fully human. Many societies, including those of ancient China and ancient Greece, refer to themselves with terms that are associated with full humanity, whereas they refer to people from alien cultures and distant lands with terms that lack that association. For example, as the classicist G. E. R. Lloyd has reminded us, those called *barbarians* by the ancient Greeks were people whom the Greeks perceived or mocked as not speaking a fully human language, merely "ba-ba"-ing instead like animals. Thus, the idea of there being different sorts of people across time and space, not all of whom are valued equally, is many thousands of years old. Given the differential valuation of different sorts of people, we have marked variation and associated differential normativity, where some people (e.g., the Greeks) see themselves as the norm, and others (those they call *barbarians*) as subnormal.

This differential normativity and the associated idea of subnormalcy played out with respect to a variety of disabilities in the ancient world. In her work on deafness and disability in the ancient world, Martha Edwards makes it clear that deaf and mute people were seen as defective sorts of people, lacking and diminished with respect to intelligence and reasoning, just as those manifesting physical disabilities were subhumanized. The social construction of disability did not begin in the nineteenth-century eugenic moment, as Davis implies, because the notions of normativity and normalcy that inform that social construction are much more deeply ingrained than he supposes.

Recall also the chief point that was made in section 5.4, namely, that normalcy and normativity play a role in all three aspects to disablement that we have distinguished: impairment, disability, and handicap. Because disablement itself, as opposed simply to some aspect of disablement, for example, handicapping, is immersed in normalcy and normativity, to view normalcy as a relatively recent, societal-specific development is to limit the range of phenomena that fall under the umbrella of disablement. In *Enforcing Normalcy*, Davis himself characterizes one of the points of the consciousness-raising about disablement done by disability studies to be "to reverse the hegemony of the normal and to institute alternative ways of thinking of the abnormal" (49). Yet by advocating a claim about the history of normalcy that locates its origins at the beginning of the short history of eugenics, the social constructivist response effectively curtails where one can identify disablement. If the disabled body really *is* invented through human nature suggests that there would be much work to be done to grate that response with our prosocial nature. Likewise with respect tc third desideratum, since it appears that the social constructivist vie\ disability will simply be accompanied by other constructivist views of r ethnicity, and other social categories for which the puzzle of marked vɑ tion arises. And since the appeal to biopolitics is silent about the phenc enology of marked variation, that appeal also falls short with respect to fourth desideratum.

There are other respects in which views focused on the nineteent century social construction of the normal are more limited, and may ev\ be thought to be misleading, when it comes to how we should think abo\ thinking about human variation, subnormalcy, and disablement. For exan ple, although such views can recognize marked variation as an ongoin concern—a kind of downstream effect of the social construction of the nor mal in the nineteenth century—they are less well positioned to identify that concern *prior to* the explicitly eugenic era. If there is reason to think that the puzzle of marked variation not only postdates but also predates the explicit eugenic era, then this is a more serious problem for the appeal to biopolitics.

Here we can take up a point that Davis himself stresses: that there is a close connection between normalcy, norms, and normativity. Because of this connection, and because the norms and normativity that are deeply entwined with marked human variation concern, in part, physical and psychological structure and function, normalcy itself is a notion that plays a role in thought about that variation in a wide variety of cultural contexts. When Aristotle says, in Book VII of *The Politics* that "as for the exposure and rearing of children, let there be a law that no deformed child shall live"—a quote that Davis knows since it forms the epithet to *Enforcing Normalcy*—I take some kind of cluster of normative and normalizing notions to be in play. For Aristotle's view of deformed children was part of a broader, socially accepted view of a variety of disabilities in the ancient world which often operated via the idea of there being distinct sorts of people. Some of those sorts of people—deformed children, for example—did not simply fail to attain some ideal, but were *subnormal*, sufficiently so that "there be a law that no deformed child shall live."

Thinking about members of our species in terms of various sorts or kinds can be found in ancient civilizations and is often bound up with the very eugenics, then there are no disabled bodies to identify prior to that point in history.

In effect, by homing in on the social construction of disability in the heyday of eugenics, the appeal to biopolitics is intrinsically limited in making sense of marked variation prior to the origins of eugenics. This is to suggest one way in which the appeal to biopolitics is *shallow*: it operates with an explanatory toolkit that cannot adequately address the task at hand. We might bring out this shallowness by articulating two related problems that the appeal has, one concerning what is missing from that toolkit, the other concerning whether social constructivist accounts really grapple with the puzzle at all.

The shortcomings of the constructivist view with respect to our list of desiderata provide some idea of what that view is missing. Each of those four desiderata suggests that there is something underlying the specific social construction of disability built through nineteenth-century eugenics, and at least two of them—the compatibility with our prosocial nature and accounting for the phenomenology of disablement—most naturally suggest that this something is *psychological* in nature, something concerning how we function as psychological creatures enmeshed in complicated, socially constructed worlds.

The final respect in which the appeal to biopolitics might be thought to be explanatorily shallow is that it doesn't solve the puzzle of marked variation so much as simply push that puzzle back one step further. For in positing that the ideas of normalcy and subnormalcy originate in a nineteenth-century eugenic nexus, this response invites the obvious question about the origins of *that nexus*. If marked variation originates in the ideas and practices that constituted the beginnings of eugenics, *where did those ideas and practices originate?*[6]

5.7   Constructivism's Open Question and Further Desiderata

For a proponent of the constructivist response to the puzzle of marked variation, this remains not simply a further question but a question that suggests that this answer doesn't get at the heart of the puzzle of marked variation. One way to bring this out is to see this remaining question as an open question in the sense articulated most notoriously—at least among the analytic *cognoscenti*—by the early twentieth-century philosopher G. E.

Moore in his discussion of ethical naturalism. Naturalistic accounts of the nature of ethics, popular during Moore's time and whose best-known examples came from utilitarian approaches to morality, analyzed moral properties, such as being right or good, in terms of nonmoral properties, such as happiness or pleasure. For example, a traditional utilitarian view is that an action is right just if it maximizes the level of happiness among those affected by the action. Moore thought that such analyses of ethics could not be correct, and his open question argument was one way to show why. He argued that for *any* such naturalistic account, it is possible to coherently wonder whether the moral property being analyzed was in fact identical to that natural property, taking that possibility to indicate the falsity of the proposed account. It was, Moore suggested, an *open question* whether any such proposed analysis was correct, and he used that suggestion to argue against naturalistic views of ethics in general.

While I think that Moore's argument itself is problematic in many ways, particularly in this final inference, what it successfully points to is the idea that naturalistic analyses leave out something critical about morality, pushing consideration of that feature one step further back in the chain of inquiry. What naturalistic views fail to capture is the *normative force* of moral ascriptions, their connection to action-guidingness and moral motivation. Likewise, the appeal to biopolitics at the core of the social constructivist response to the puzzle of marked variation omits something fundamental. Its answer to the determinate form of the question "Where do ideas of human variation come from?" that has been the focus of this chapter is subject to the kind of further investigation that the open question argument relies on. So although an appeal to the kind of biopolitics that emerged in the second half of the nineteenth century, one in which science and medicine play a role in disciplining disablement, takes us some way to understanding the general question "Where do ideas about human variation come from?," this kind of social constructivism leaves us with a truncated answer to that question. That truncation expresses a fundamental limitation to understanding the puzzle of marked variation through an appeal to a socially constructed eugenic moment.

We can integrate this limitation of constructivism with our previous diagnosis of its shortcomings with respect to our four desiderata to arrive at an extended list of desiderata for any adequate response to the puzzle of marked variation. In addition to the four desiderata we have already recounted, an adequate response to the puzzle should:

5. Allow one to make sense of disability prior to the explicit eugenic era;
6. Connect the social and psychological dimensions to the puzzle of marked variation or, alternatively, explain why one of these is sufficient as a response to the puzzle;
7. Solve or bypass a regress, one that simply shifts back the puzzle of marked variation to the next level of the *explanans*.[7]

5.8 Conclusion

I have suggested that the general question "Where do ideas of human variation come from?" can profitably be addressed in thinking about disability and eugenics in conjunction with the more specific (and somewhat more cumbersome) question "Where do ideas of normalcy and subnormalcy that mark disablement come from?" I have challenged views with a certain currency in disability studies that draw on Foucault's notion of biopolitics in sketching a view of normalcy and subnormalcy. Such views make these notions relatively recent social constructions that emerged with eugenics, and I have focused my critical discussion of this idea on the influential work of Lennard Davis. The kind of subnormalcy-marked variation found in medicine and the health sciences has changed over time, taking a specific form in the late nineteenth century with the rise of eugenics. Yet the puzzle of marked variation runs deeper than an appeal to this eugenic moment can address. For all the popularity of appeals to biopolitics in understanding the nature and history of eugenics, those appeals reach a limit in the response that they provide to the puzzle of marked variation. At best, they form part of an answer to the question "Where do ideas of human variation come from?" and we need to explore other kinds of answers.

The other kind of answer that I shall develop in the next chapter will delve further into underlying psychological tendencies that structure the human psyche, grappling with the kind of socio-cognitive beast *Homo sapiens* is. Given the preceding desiderata, this appeal to our underlying human nature will not necessarily aim to reduce "the social" to "the psychological." It also need not view disablement as a natural or innate category of the mind, nor accept that there is something immutable about thinking about variation that makes disablement an inevitable consequence. Rather, it is an attempt to open up a way of thinking about our thinking about human variation that recognizes the relationships between certain features of our social circumstances, the role of norms and normalcy in regulating social structure, and the place that cognition plays in this.

The ways in which the socio-cognitive framework for addressing the puzzle of marked variation supplements—and the ways in which it conflicts with—appeals to biopolitics here is something to return to once both views are before us. Each might tell us something important about our thinking about these questions about human variation, and create some space for further questions about disability, medicine, and the health sciences.

6 A Socio-cognitive Framework for Marked Variation

6.1 A Hobbesian Prelude: Born to Be Not-So-Wild

Given the framing of eugenics against the backdrop of our prosocial nature that I introduced in chapter 1 and elaborated on in chapter 5, one might expect that I would take the seventeenth-century English political philosopher Thomas Hobbes to be just another Bad Guy in the History of Ideas, a kind of antisocial social philosopher. After all, Hobbes famously depicted human life without any form of government—life in what he called the *state of nature*—in unflattering and pessimistic terms. Without the security afforded by state power, thought Hobbes, we would be in a war of "each against the other," and our lives would be "solitary, poore, nasty, brutish, and short." Hobbes's views about this state of nature, human nature, and sociality here have been, and remain, central to the tradition of social contract theory within liberal political theory.

And those Hobbesian views have been taken typically to rest on the assumption that human nature is ruthlessly self-interested, and to imply that we are sociable only for instrumental reasons of self-advantage. Hobbes is the quintessential political philosopher of selfishness, the philosopher of what the Canadian political philosopher C. B. Macpherson has called "possessive individualism." On this standard view of Hobbes, his views of the fictions of the state of nature and the social contract through which we enter political civilization make Hobbes a defender of the view that we are so constituted as a species that we must accept a form of government with absolute sovereignty, a "Leviathan" against which we have no natural rights as human beings or as citizens.

I think this standard interpretation of Hobbes is badly mistaken, however, and that Hobbes in fact has a prosocial view of human nature much closer to the one I have sketched. In fact, I take Hobbes to share the Aristotelian view that we are by nature sociable creatures, albeit ones who could readily find ourselves in circumstances that logically demand deference to state (or other) authority and require abandoning what Hobbes called "private judgment." But even putting this aspect to Hobbes's views on human nature to one side, there is something more immediate in his views that make him an ally rather than a foe in debates about human prosociality.

The characterization Hobbes gives of human life in the state of nature—solitary, poor, nasty, brutish, and short—does accurately picture what each of our lives would be like from infancy, not in the absence of the protective presence of the state, but without the facilitative presence of the care-givers that each and every one of us needs to survive that natural condition. What Hobbes notes about our vulnerability in the state of nature—that even the strongest of us can be killed by the weakest when the strongest is asleep—applies to each of us more generally during the natural states of infancy and early childhood. In short, our continued survival beyond early life depends minimally on others not knocking us off. More generally, without being able to rely on more active forms of basic assistance and aid—food, shelter, protection, removal from harm's way—the life of each of us would be brutish and short, whether or not solitary, poor, and nasty. But we survive. Thus, we must be a prosocial species, even if imperfectly so.[1]

6.2   Sociality and Prosociality

I have returned to this variant on the argument for our prosocial nature because the form that our prosociality takes is important for the response that I would like to develop to the puzzle of marked variation. Here we need to take a step back and reflect on the kind of sociality we have. That first step involves identifying what it means to be a social species, or to have sociality as a core feature of our nature.

We are a social species, but not uniquely so. For starters, we share many of our constituent forms of sociality with the two hundred or so species in the Linnaean order that we fall under, the primates. Sociality is part of our primate heritage, with various social traits that we possess being phylogenetically derived from the nearest common ancestors we share with existing primate species. And sociality, in even more various forms, is found across phylogenetically quite distant parts of the living world: in mammals and in so-called social insects, for example.

In the mobile living world, sociality is pervasive, and it is easy to understand why. Living agents that move around need to have means of responding (relatively) rapidly to features of their local environments that can change (relatively) quickly because of the movement of the agent. This is why sensory systems are ubiquitous in the mobile living world. When you are moving around, your proximal environment tends to change more rapidly in ways that are relevant to your survival and reproduction than it does when you have a sedentary way of life. This is why mice have elaborate, quick-time sensorimotor systems, but trees do not.

Any mobile living agent, unless it is extremely unfortunate or unusual, will often encounter other mobile living agents that are endowed with something like the same capacities and powers that it has, in part because it will be reproduced by, and often with, other such agents. For a mobile living agent to track, respond to, and even anticipate the behavior of other mobile agents requires even more sensory or cognitive sophistication than simply to track, respond to, and even anticipate other kinds of environmental resources.

Social interactions in the nonhuman animal world take many overlapping forms. They can be, among other things, reproductive, cooperative, competitive, predatory, protective, domineering, resource-securing, mutualistic, exploitative, parasitic, pathogenic, altruistic, or sacrificial. Many of these forms of sociality take place with very little cognitive mediation, given that they occur between critters whose individual cognitive power is likely quite limited. Much of this sociality is merely aggregative in that it is the outcome simply of co-occurring individual behaviors that require little coordination with conspecifics. Jellyfish, like real fish, tend to form social aggregates; these jellyfish blooms, unlike the fish schools they may superficially resemble, fall into this category. Because at least much socially aggregated behavior can be done with minimal attention to conspecific behavior, it imposes a relatively low cognitive demand on the individual.[2]

6.3 Human Sociality and Its Cognitive Demands

Much socially coordinated behavior in both the human and nonhuman animal world, however, goes beyond simple forms of social aggregation.

When it does, the cognitive demands of that coordination can be considerable. There are three general strategies for meeting those demands, all of which have been adopted in *Homo sapiens* extensively in shaping up the unique forms of sociality that we possess.

The first is to develop or acquire more sophisticated internal cognitive processing. This might include developing special sensitivity to cues that allow one to detect other nearby mobile creatures, or the specific state that they are in, indicative of whether they are friend or foe, potential mate or likely predator. It might include evolving special memory systems that allow the organism to recognize particular individuals, or even to track their past behavior. Or it might include building reflex-like responses to others, motor routines that allow rapid reactions in the face of expected or unexpected social behavior from these others. The most influential explorations of the cognitive demands of distinctive human sociality have focused on the development of such internal cognitive sophistication as representing the key evolutionary changes in our lineage that facilitate distinctive forms of socially coordinated behavior.

For example, the much-discussed *Machiavellian intelligence hypothesis* holds that the evolution of human intelligence was driven primarily by the cognitive demands of increasingly complex social coordination problems. The development of a theory of mind, allowing for an understanding of one another in terms of internal and sometimes nonmanifest mental states whose representational content can depart from that of what they represent, has been taken to be a significant cognitive achievement here. The basic idea here is that human sociality is both made possible and facilitated by increasingly complex internal computational processing—processing that is highly modular in response to specific selection pressures.

The second general strategy for meeting the cognitive demands of human sociality is to distribute the requisite sophistication to one's cognitive processing between internal and external resources, in effect making use of various forms of situated, distributed, and extended cognition. While the general idea of distributed cognition has long found a home within the cognitive sciences, it has only been more recently that the relationships between distributed and extended cognition have been explored. Just as the idea of embodied cognition posits the distribution of cognitive loads between brain and body, the hypothesis of extended cognition proposes that this distribution extends beyond the body.

The reliance of cognizers on external storage systems—parts of the physical world that allow us to extend our biological cognitive capacities by storing information in the world and accessing that information as needed—is one primary way in which extended, situated cognition operates. But there are now philosophical accounts on offer of a variety of cognitive processes—for example, visual perception, moral cognition, emotional development, and musical cognition—that begin to round out the scope of extended cognition. On the *extended mind hypothesis*, parts of the world can be physically constitutive of cognition, not only as external storage systems, but also more generally. Cognition involves the incorporation and appropriation of *cognitive resources*, and the intuitive idea behind the extended mind thesis is that the resulting cognitive extensions can involve resources wherever they are located vis-à-vis the physical boundary of the organism itself.

The third strategy is not to distribute individual cognition through the adoption of situated and extended forms of cognition, but to supplement or even replace individual cognition with some kind of collective or group-level cognition. To fix on what *group minds* or *collective cognition* might be, consider the so-called "social insects," the *hymenoptera*—the wasps, ants and bees—together with the termites.

As their name suggests, the social insects exhibit much sociality, from nest-cleaning behavior to hive temperature regulation requiring the coordination of the behaviors of thousands of bees. Although individual wasps, ants, and bees clearly have some level of cognition, on an individual level it is relatively limited. Despite this, social insect colonies as a whole or in sizeable part, accomplish impressive outcomes that are very naturally described by attributing perceptual or sensorimotor properties to those groups of organisms. These include the perceptual and communicative abilities involved in gathering information about food sources and the motoric capacities to utilize resources and avoid predators and dangers in the world. Some of these capacities or abilities manifest both some level of intentionality and a degree of concern over the integrity of the colony. These include the ability of a bee colony to locate distant sources of nectar and regulate the relative number of foragers and hiver workers in accord with the richness of the source, and the ability of a termite colony to rapidly repair damage to its nest. Yet it is very implausible to think that these abilities are possessed by *individual* members of the hive, nest, or colony. In short, the behavior of at least some groups is such that it seems directed at self-preservation, where the self here is a colony, and the means of achieving that goal involves group-level decision making that draws on collectively distributed perception and sensing. The relevant, putative cognitive activities here—for example, perceiving, remembering, deciding, monitoring—are what I have elsewhere called *group-only* traits, which are possessed only by a group and not by the individuals that comprise the group.

This appeal to group-level cognition is familiar to those working on sociality in the fragile sciences, having been made by the sociologist Emile Durkheim in his influential invocation of collective representations. Durkheim originally drew a distinction between collective and individual representation as part of an argument for the autonomy of the social sciences from psychology. Durkheim's basic claim here was that just as psychology worked with individual representations, so too would sociology be the science that studied collective representations. Contributing to what I have elsewhere called the collective psychology tradition in hypothesizing group minds, Durkheim's appeal to collective representations was both a response to perceived concerns about individual and national degeneracy and also manifested a more optimistic tendency among both psychologists and social scientists who saw in the group mind something meliorative and uplifting. For Durkheim in particular, it was through collective consciousness and representations that virtues such as cohesiveness and solidarity that offset and counteracted social factors causing individual detriment could be emphasized.

The influence of Durkheim's views here within the fragile sciences, and the reach of the collective psychology tradition into the popular imagination, have created a comfort with the idea of group-level cognition in the social sciences that has fueled recent work on collective memory, decision making, and cognition more generally. That idea has also received more intense philosophical scrutiny in recent work on shared intentionality and collective social action.[3]

6.4 Shared Intentionality and Collective Social Action

In recent philosophical and psychological work focused on human sociality, the idea of shared or collective intentionality has loomed large. Collective intentionality has come to be conceived as some kind of crowning achievement of our species, and perhaps of our closest ancestors and living relatives. It is a sort of keystone accomplishment that brings in its wake new forms of sociality, building on and utilizing forms of individual cognition that are themselves distinctively human. More specifically, over the past twenty years shared or collective intentionality has been used to explore supposedly distinct forms of human cooperation and conflict, the role of institutions in human social life, and even the broader nature of social reality itself. Precisely where (and how) we locate shared intentionality among the individual enhancements, social distribution, and group-level implementation of cognitive sophistication remains a matter of ongoing debate. While we need not enter into that issue here, it will pay to have some understanding of what motivates the appeal to shared intentionality, since it is shared intentionality that provides the basis for the *normative dimension* to human sociality important to the puzzle of marked variation.

Much human social behavior is cooperative, shared, or joint. We do things *together*: we work and play, we walk and talk, we celebrate and mourn, we laugh and cry. There seems to be little reluctance to view ourselves as undertaking such behaviors or actions together, to accept collective action, in addition to individual action. Collective actions, such as building a fire together or holding hands, are no more ontologically dubious than the corresponding individual actions. Consider *distributive* and *joint* or *shared* forms of collectively acting.

A collective or group of individuals acts *distributively* when the components of the overall action are distributed across the actions of those individuals. In distributive collective action, the group does something that no individual in the group herself does, except insofar as she contributes to the collective action itself. To take a simplified example, one beaver finds and transports waterlogged debris to a particular place in a creek; a second beaver then places that material in the growing dam. The collective action of building a shelter—a beaver lodge or dam—is distributed across this pair of actions. To find and transport that material, or to arrange it, is to build a shelter only insofar as these component actions form part of the collective action.

For there to be *joint* or *shared* collective action, there needs to be not simply distributive collective action but in addition some kind of coordinating glue that makes it an action that is completed *together*. When a team of contractors builds my house, or a restaurant cooks my meal, there is not simply distributed collective action but also the kind of coordination and cooperation that makes for joint or shared collective action. What is the coordinating glue in such an action? One hypothesis is that joint or shared *intentionality*, particularly shared *intentions,* is what provides this coordinating glue. Such shared intentions have been central to the literature on collective intentionality, where they are often called "we-intentions": first-person plural intentions. But the more important point here is that this coordinating glue is *something mental,* however it is conceived more precisely.[4]

6.5   Sketching the Socio-cognitive Framework

So my starting point is a certain view of sociality in general (section 6.3), and of human sociality in particular (section 6.4). I take sociality to be a ubiquitous feature of the biological world, especially the mobile biological world containing creatures with some self-governing capacity to move from location to location. Sociality requires some kind of at least minimal group living, interactions with conspecifics, and a differential sensitivity of some kind to the presence of others, both conspecifics and nonconspecifics. While sociality per se does not mark us off from other species, the specific form that our sociality takes does. I hypothesize that human sociality is distinguished by three related features:

- Differential sensitivity to others is *cognitively* mediated in ways that go beyond enhancement of sensory systems.
- The resulting sophistication to our sociality rests on a variety of forms of individual, extended, and collective cognition.
- The corresponding cognitive mechanisms are enmeshed with some kind of *normativity* that is generated by shared intentionality and that, in turn, regulates the forms that our sociality takes.

Having already introduced the key component ideas here in the previous sections, I will simply say something briefly about each of these three features.

Cognitively Mediated
Not all mobile organisms that display a differential sensitivity to others have cognitive capacities, and even when they do have such capacities,

their reactions to others are not always governed by their cognitive mechanisms, but by chemical sensitivities (in the case of bacteria and other single-celled organisms, as well as social insects); by the detection of threshold levels of environmental resources (in insects and birds); and by low-level perceptual mechanisms that do not feed into higher cognitive capacities (in birds and mammals).

Spelling out both just what makes a process *cognitive*, and precisely when we find cognition proper in the natural (or even artificial) world, are notoriously thorny issues. But some standard answers to these questions are helpful in corralling the phenomenon of cognition. Cognition involves *representation crunching*, and the processes that crunch those representations are often *computational* in nature, ideas at the heart of the classic computational and representational theories of mind, but also available to proponents of alternatives to those views (e.g., connectionism). Insofar as cognition proper goes beyond both perception (on the input side) and behavioral reflexes (on the output side), it involves interactions between internal representations, as well as between internal and external representations, that give rise to the agility and flexibility in behavior that is the beacon of underlying cognitive processing. Cognition is not uniquely human, but it is also not ubiquitous in organismic responses to the social world.

Sophistication of Sociality through Individual, Extended, and Collective Cognition

Our response to evolutionary pressures and social circumstances includes not simply more complicated internal processing but also forms of cognition that are distributed among mind, body, and world, as well as those that operate at the level of the group itself. Extended cognition allows us to create cultural capital that serves as common cognitive resources—as when we devise writing systems—and then to make use of such common cognitive resources for individual cognitive extension—as when we use pen and paper to jot down notes for ourselves. While I think that nonhuman animals have made use of extended and collective cognitive systems that rely on shared intentionality, human animals have made themselves masters of these cognitive trades. Traditions, rituals, ceremonies, rehearsals, and cultural symbols are among components of extended cognitive systems familiar to those in the fragile sciences of sociality, particularly cultural anthropologists and sociologists.

Normativity via Shared Intentionality

Normativity exists when there is a distinction between a correct, proper, or appropriate way for a process, event, or outcome to turn out, and an incorrect, improper, or inappropriate such way. Like extended cognition, normativity arises in and through both nonhuman and human cognition; it is not solely a feature of our own species' activity. But again like extended cognition, the most familiar and robust forms of normativity are those that are the product of distinctly human practices and institutions that presuppose a kind of shared intentionality. This normativity encompasses the norms generated within legal systems, within codes of etiquette, and by morality. Such norms may be made explicit in the form of rules or commands, or may be implicit in the ways in which we interact with one another. The threefold sophistication to our cognitive processing structures our sociality partly through this normativity.

So we have a kind of externally mediated, cognitively driven normativity, and it constitutes an important feature of human social life. One thing that this cognitively mediated normativity does is allow us to distinguish not simply between individual people but between *kinds* or *sorts* of people. Because of this dimension to our cognitive toolkit, there is an important place for the appeal to human cognition in addressing the puzzle of marked variation.[5]

6.6   Sorts of People, Normativity, and Marked Variation

In saying that we distinguish between kinds or sorts of people I mean neither that we ordinary folk have rich conceptions of natural kinds that we apply to people, nor that human kinds or sorts *are* true natural kinds existing in the world independently of how we think about the world. (For these reasons I tend to speak of *sorts of people* rather than *kinds of people*.) Rather, in saying that we distinguish between sorts of people I mean only that we employ various categories to distinguish among those whom we recognize as people. In fact, we do so quasi-recursively: once we have already sorted people by one type of criterion, we then continue the process with other subsidiary criteria. People sort one another (and themselves) into many different categories: by their height and weight, their eye, hair, and skin color, their sex and sexual behavior, their income level and type of employment, their personality and beliefs, their tastes in recreation and entertainment, their ancestry, religion, and ethnicity, their astrological sign and year of birth, and their marital and parental status. This mixture of cross-cutting and hierarchical sorting is a ubiquitous feature of how we think of one another and ourselves. Yet such sorting is not a mere mental exercise, for it contributes both to what one does, and to the normative constraints that apply to oneself and to others.

One of the primary ways we use such human sorts is in determining whether other people are, in some important way, *like us*. The plural referent "us" in "like us" is deliberate, and refers to some type of social group. Therefore, in saying that we sort people in terms of their being—or not being—like us, I am saying that we sort them in terms of whether they belong to our social group. Social groups, as I am thinking of them here, are simply groups of people. They can vary in terms of their permanence, in their value to how we lead our lives, in their relationship to our own sense of identity—of who *we* are—and in the extent to which membership in them is up to each of us, what we might think of as a dimension of Heideggerian thrownness to belonging. Others who are "like us" might share any of our appearance, our sex, our skin color, our ancestry, our language, our history, our values, our social position, our interests, our political engagements, or our activities. When they do so, they are "one of us," along one or more dimensions. When Meryl joins the Baker Street Tennis club, she becomes one of us (if we are already club members); when Bryan takes out Canadian citizenship, he becomes one of us (if we are already Canadians); and when a child is born in contemporary Western societies, he or she becomes *one of the family* (again, if that child is *our child*).

This sorting of human beings is no mere matter of cognitive registration. It has normative uptake, often of a significant kind. It also generates a distinctive type of knowledge, departing from typical third-person knowledge, knowledge of facts that could be known by anyone. Instead, it generates information about how others relate to oneself. As such, it is properly conceived as a socially located agent, first-person plural knowledge: *first-person* because it is knowledge *of one's self*; *plural* because it is not just me-knowledge but *we-knowledge*, knowledge that I have of myself as a part of some *us*. Much of this first-person plural knowledge is fleeting, but it is that which is lasting that is of interest here.

The connection between externally mediated, cognitively driven normativity and the "one of us" deployment of human sorts is something like this. Prominent among the norms that feature in our social lives are those that are generated by, and form a part of, the social groups that we belong to, including those that are significant for our identities. These are *our norms*, and not only do they apply to those in our group, but they also are enforced and transmitted through the extended cognitive systems utilized by that group. Like human sorts themselves, norms can vary from the frivolous to the identity-determining, as well as along other dimensions of variation. Our externally mediated, cognitively driven normativity is group focused, and it is this that provides the link to our employment of human sorts for the determination of co-belonging. For we use human sorts in this way as a means of engaging in the kind of externally mediated, cognitively driven normativity that characterizes and pervades human sociality.

Suppose that we do have "like us" or "one of us" human-sorting mechanisms. And suppose that disablement—like race/ethnicity and sex/gender—is one complex domain that these mechanisms operate within. Then all it would take for us to end up with the kind of difference between people being marked as subnormal is for the norms that make someone *not like me*, that is, not a member of my group, to be ones that class as subnormal those people who have (or even simply are perceived to have) disabilities or impaired parts. Such norms are abundant, both in our shared social and cultural spaces and in the internal processing that mediates individual cognition.

We can express this point in terms of the idea of *cognitive scaffolding* that is commonly invoked in work on distributed and extended cognition. All societies and cultures contain a lot of external cognitive scaffolding that stigmatizes disablement in ways that direct and reinforce our "like us" mechanisms. What the historical eugenics movement provided were specific ways to think about human improvement that centered on the elimination of disability, particularly cognitive disability, by making the category of feeble-mindedness central to the conception of a range of social ills and how they could be cured in future generations. By introducing cognitive scaffolds that subhumanized the feeble-minded, but also others whose questionable mental health was a key eugenic trait, eugenics channeled our cognitively mediated sociality in a particular way, both building on existing dispositions to cognitively react to one another via "like us" detectors and continuing to structure how people with disabilities, especially intellectual ones, are viewed in our supposedly post-eugenic society.

Thus, the ideas of normalcy and subnormalcy that mark disablement and medical pathologization derive from a kind of cognitively mediated normativity that is created, reinforced, and transmitted through individual, extended, and group-level cognition. These ideas are thus rooted in distinctively human forms of cognition and sociability, and understanding them more fully involves further specification of mechanisms that operate across the cognitive and social domains. I have speculated that "like us" detectors constitute one such mechanism. On this view, our ideas about marked human variation are neither distinctively nineteenth-century nor intrinsically tied to what is often thought of as the heyday of eugenic thought and practice. Those ideas both predate and postdate the eighty-year period occupied by the short history of eugenics (1865–1945), and likely underpin ongoing views of, and attitudes toward, human variation and disability. Independent of the kind of social constructivism about normalcy that took place in the nineteenth century, there is a further socio-cognitive dimension to our responses to marked variation associated with the detection of subnormalcy among human variation.

6.7 Clarifying What First-Person Plural Mechanisms Are

In articulating a socio-cognitive framework for addressing the puzzle of marked variation I have spoken loosely of "one of us" and "like us" mechanisms, and suggested that these are first-person plural mechanisms. In this section I want to sharpen my characterization of such mechanisms and the corresponding first-person plural knowledge they generate by considering two much-discussed hypotheses about human social cognition, the first arising within anthropology, the second in developmental psychology.

In "Are Ethnic Groups Biological 'Species' to the Human Brain? Essentialism in Our Cognition of Some Social Categories," the anthropologist Francisco Gil-White brought the adaptationist thinking of evolutionary psychology and evolutionary anthropology to bear on the question of how we perceive ethnic groups. Dissatisfied with the predominance of constructivist perspectives on the anthropology of ethnicity, and reintroducing the idea of essentialism back into thought about ethnicity, Gil-White argued for a particular hypothesis about how ethnic groups are perceived. Here Gil-White invoked a specific mechanism, a module for processing *living kinds* that was familiar in literature on cognitive development. In keeping with the rejection of traditional essentialism that is widespread in both the biological and social sciences, Gil-White acknowledged that ethnic groups do not in fact have essences: there is no set of intrinsic properties, each necessary and together sufficient, defining membership in an ethnic group. Yet attuned to the developmental literature on *psychological* essentialism, Gil-White nonetheless wondered aloud about why it is that ordinary people often treat ethnic groups *as if* they had essences. His core hypothesis, as summarized in the abstract to his paper, is as follows: "Humans process ethnic groups (and a few other related social categories) as if they were 'species' because their surface similarities to species make them inputs to the 'living kinds' mental module that initially evolved to process species-level categories."

Gil-White's argument is something like this. We have a "living kinds" module, which is a cognitive adaptation that evolved to detect species. Ethnic groups on the surface are similar to species, and for this reason, we treat them cognitively as if they were in fact species. Thus, ethnic groups serve as a trigger for our living kinds module, which is why we treat ethnic groups as if they had essences, when in fact they do not. Put in terms that the philosopher and anthropologist Dan Sperber has introduced, we have a module whose *proper* domain is species-level categories, but whose *actual* domain is ethnic groups.

Gil-White's hypothesis posits a psychological mechanism, a living kinds module, that detects in the biological world a certain natural kind: species. Gil-White considers species themselves, of which *Homo sapiens* is one example, to have intrinsic essences. Our living kinds module is also pressed into service in our perception of a certain kind of social group, an ethnic group. We can (and do) come to distinguish between our own ethnic group and those of others. But there is nothing about the living kinds module that makes the comparison between ourselves and others crucial to its operation. It is in this sense that the living kinds module generates *third-person* knowledge.

On Gil-White's view, our evolutionary history is one in which we first have a living kinds module for detecting species in general, applying it to our own case to detect conspecifics and distinguish ourselves as members of a common species from other organisms. We then come to use this module in this same way within our own species to detect people who are co-ethnic and distinguish ourselves as such from individuals in other ethnic groups. The shift here is from third-person to first-person knowledge. But rather than starting with species detection, might we not start with a cognitive endowment that allows one to detect those like oneself in some salient way—in perceptual appearance, in behavior, in language, in where we live—delivering knowledge that is *first-person* in that the content it delivers is inelimibably tied to the perspective of the particular cognizer? On this view, our evolutionary history is one that starts with a psychological mechanism that generates first-person plural knowledge—knowledge about us—with that knowledge being later, if at all, generalized to the level of conspecifics and to the species-level more generally. The evolutionary trajectory here, in short, would be just the opposite to that implied by Gil-White's hypothesis.

Cognitive mechanisms that are first-person rather than third-person in nature have been central to the work of the developmental psychologist Andrew Meltzoff on imitation and sociality. Building on a now classic study that showed that very young infants spontaneously imitated an adult movement, such as a tongue protrusion, over the past forty years Meltzoff has articulated a theoretical framework for integrating subsequent findings about imitation, gaze direction, mental state detection and attribution, action, observational learning, and first-person experience in infants under two years of age.

The basic idea to Meltzoff's theory is that infants use first-person knowledge of their own intentional actions in order to understand the actions of others and the unobservable mental states—goals, desires, aims—that generate them. Meltzoff is concerned with presenting a view of infants that overturns the dominant view of them as asocial creatures, the sort of influential view that one finds in thinkers as different as Freud and Piaget, but does not rest content with the claim that we are born with an innate theory of mind that simply unfolds over time. Infants are geared to imitate *actions*, where these are not simply movements but behaviors that are generated by intentions. Crucially, they can use their own bodily self-knowledge in order not only to imitate but also to anticipate what agents will do, and what it is that those agents want to happen from their intentional actions. While some bodily control (and the self-knowledge that it presupposes) is an arduous and late (if ever) acquired accomplishment, more important here is the fact that both can take forms that are immediate and effortless, and can even involve little cognitive effort. Tongue protrusion, bodily orientation, head turning, gross reaching, and vocalization are all bodily actions that even very young infants can engage in with intentional purpose, and in interaction with another person. They presuppose some bodily self-knowledge, and their recognition in others—also manifest in young infants who are subsequently motivated to engage in the *same* actions—presupposes a mechanism that generates first-person plural knowledge: what Meltzoff calls "like me" detectors.

Although these mechanisms do generate some first-person plural knowledge, their principal outputs are first-person *singular* knowledge, knowledge about one's *own* actions, and how it coordinates, compares, and differs with those of others in one's immediate environment. By contrast, for the mechanism that I have in mind as underpinning our sense of group identity, the "like us" mechanism, first-person plural knowledge is an intrinsic output: the "like us" mechanism delivers representations that are not only comparative in nature, but compare individuals vis-à-vis the social groups they are perceived as belonging to. Thus, the postulated "like us" mechanism tells those who possess it not simply whether another is the same as or different from oneself, as does Meltzoff's "like me" detector, but by doing so with respect to social group membership carries information about the corresponding social groups.[6]

6.8   Return of the Seven

Both the discussion of the constructivist view that appeals to biopolitics from chapter 5 and that of the socio-cognitive framework introduced in this chapter invite further questions and call for more details. Yet I hope that the general contrast between the two views should be clear in terms of the response each gives to the puzzle of marked variation. The constructivist view sees marked variation as closely tied to the rise of eugenic ideas and practices. By contrast, the socio-cognitive framework takes the eugenic nexus to provide just one particular way in which a recurrent socio-cognitive disposition plays out.

In terms of the seven desiderata for any adequate response to the puzzle of marked variation, I have argued that the constructivist view's only real strength lies in the first desideratum. Recall that these say that that response should:

1. Recognize the puzzle of marked variation as an ongoing concern, rather than one arising only for an explicitly eugenic past;
2. Ensure that the account of marked variation it provides is compatible with our general prosocial nature as a species, and explain how the two are to be integrated;
3. Explain how various manifestations of the puzzle of marked variation—those arising with respect to race, ethnicity, and disability, for example—are related to each other;
4. Account for the felt immediacy to the phenomenology of marked variation, addressing both dimensions to this felt immediacy, the one concerning perceptual input, the other concerning emotional output;
5. Allow one to make sense of disability prior to the explicit eugenic era;
6. Connect the social and psychological dimensions to the puzzle of marked variation or, alternatively, explain why one of these is sufficient as a response to the puzzle;
7. Solve or bypass a regress, one that simply shifts back the puzzle of marked variation to the next level of the *explanans*.

I think that the socio-cognitive framework satisfies all of these desiderata, and that it is quite easy to see why it does so.

This is chiefly because the underlying cognitive mechanisms that it posits—like us detectors—operate on normatively meaningful social categories, some of which (such as kinship) are key to our prosocial life (desideratum 2), others of which operate on marked variation in other domains (desideratum 3). With a focus on cognitive mechanisms that can operate in very different social and cultural circumstances, this view is also apt for making sense of disability both before (desideratum 5) and after (desideratum 1) the explicit eugenic era. In positing first-person plural knowledge, it is geared both to capture the phenomenology of marked variation (desideratum 4) and to connect social and psychological dimensions to the puzzle (desideratum 6). By responding to the puzzle of marked variation by appealing to our socio-cognitive nature, it avoids the particular version of the explanatory regress represented by the open question-style argument I gave (desideratum 7), though it does invite those skeptical about nonreductive, naturalistic responses to the puzzle to articulate their own form of this problem.

6.9   Standpoint Eugenics in the Socio-cognitive Framework

Grappling with the puzzle of marked variation has brought out further dimensions to eugenics, past and present, and called into question the adequacy of the dominant social constructivist narrative about disability and its relationship to eugenics. Human variation is a part of the natural world, and marked variation within that is a part of a human world structured by forms of cognitively mediated sociality. I want to complete this extended discussion of human variation and disability by taking up one further sorting of people within a category of disablement—one that brings us back to the idea of standpoint eugenics more explicitly.

Consider *thalidomiders*, people who were, and still sometimes are, born with visible variation in their limb formation as a result of their mothers having taken the drug thalidomide during pregnancy. Thalidomide, first sold under the name Contergen in Germany late in 1956 as a sedative treating morning sickness and insomnia, came to be approved in almost fifty countries over a five-year period before its adverse effects on the development of the fetus were verified and the drug withdrawn. (The drug was never approved by the Federal Drug Administration in the United States, but was available in Canada in 1961–1962.) Thalidomiders have variation in any or all of their limbs, including extreme truncation of one or more limbs, more moderately shortened limbs, and variations in digit number and location when the hand or foot is formed. Most thalidomiders in the West were born between 1958 and 1962, with the majority born in Germany, where thalidomide was sold earliest and for the longest period. The case of thalidomide was a principal motivation for more rigorous testing—especially on pregnant women—of pharmaceutical drugs before they received regulatory agency approval. It also played a broader role in shifting social attitudes in related areas, such as in attitudes toward abortion and prenatal screening.

The biochemist and disability theorist Gregor Wolbring is a German-born thalidomider and one of the biochemists who has worked on thalidomide derivatives effective in relieving some of the symptoms of leprosy and some cancers, and has drawn attention to the role of thalidomide in pharmaceutical regulation and the role of society in the lives of thalidomiders. As Wolbring puts it more bluntly, thalidomiders became the "poster children" not only for stricter regulation of pharmaceuticals, but also for the fight for a woman's right to an abortion in the era leading up to the U.S. Supreme Court decision *Roe v. Wade*. We can spell out the rationale or logic that Wolbring sees relating thalidomide-induced "birth defects" and large-scale and wide-ranging social policy changes in the Western world. To avoid something *that monstrous* being born, society should undertake to provide for the rigorous and systematic regulation of drugs taken during pregnancy; it should also ensure that any woman carrying a fetus detected as being *like that* have the right to terminate her pregnancy. Both of these societal changes were mediated via governmental policy and legislation. From Wolbring's standpoint, this provides a striking and extreme case in which the marked variation associated with thalidomiders has eerily similar life and death consequences during the last quarter of the twentieth century as did Harry Haiselden's advocacy of the euthanasia of "defective infants" in its first quarter.

Thalidomider, of course, is not a medical term, and is rarely used in medicine itself—"thalidomide victim" is the most commonly found generic term for people who underwent development changes in their bodily formation due to their mother's having taken the prescribed drug thalidomide. "Thalidomider" is, rather, a self-identifying expression, one used by such people in talking about themselves, individually and collectively, in the first-person and in the third-person. Like the general shift from "victim" to "survivor" that was pioneered within feminist discussions of child molestation, rape, and other sexual crimes, and that I have adapted here in articulating a standpoint eugenics, its introduction in the context of the extreme subhumanization experienced by thalidomiders represents a kind of reclamation of one's own personhood, dignity, and self-respect.

This self-affirming signaling of marked variation contrasts with the language of victimization, tragedy, and loss permeating the medicalized discourse that provides the dominant framing of public awareness of thalidomide-related phenomena. It represents the prosocial flipside of the one-of-us mechanisms that, I have been suggesting, structure The Eugenic Mind.[7]

7   Back Doors, Newgenics, and Eugenics Underground

7.1   Newgenics

Both the constructivist and the socio-cognitive frameworks for addressing the puzzle of marked variation recognize the contemporary significance of marked variation to the disability community, which therefore is relevant to more than the eugenic past. While both frameworks contribute to our understanding of the persistence of eugenics, I have argued that the socio-cognitive framework provides an explanatorily richer and deeper account of that persistence, and of the marked variation that underlies it. Beyond that, what marked variation is marked *for* or marked *as* are a further pair of questions whose answers affect and reflect how we view that eugenic past. The perspective of a standpoint eugenics sharpens the focus on the relationship between the eugenic past and present attitudes, values, and practices, particularly regarding disability, both by asking such questions and by providing distinctive answers to them.

As an emerging scientific enterprise originating in the last third of the nineteenth century, as well as a social movement whose heyday ended in 1945, eugenics had the aim of improving intergenerationally the quality of human lives by changing the composition of particular human populations to produce more desirable and fewer undesirable people. For many contemporary geneticists, historians, philosophers, and bioethicists, it is this idea of human improvement or betterment in the history of eugenics that looms largest in how they think of the relationship between past and present. For them, early twentieth-century eugenics was problematic primarily in its epistemic basis and execution, and they find little to object to in the idea of enhancing desirable traits found prevalently in human populations while reducing the incidence of minority traits associated with disability.

A recognition of the ill-fated nature of eugenics as a historical phenomenon requires denigrating early eugenic science and social policies, and sorting the wheat from the chaff in the eugenic past has involved recoding contemporary discourse and practices to sharply distinguish the old from the new. For example, in his defense of what he calls "liberal eugenics," Nicholas Agar proclaims that "experts on human genetics consulted by the prospective parents of tomorrow's liberal societies will give vastly better scientific advice than that given by Hitler's scientific lackeys." Enhancing humans by manipulating traits is altogether different. For bioethicists, much as for historians, eugenics *is* past. Or at least insofar as eugenics remains present in aims of melioration and human improvement to be achieved by technologically mediated and scientifically informed intergenerational interventions on human populations, it is a question of how to avoid falling into forms of eugenics that replicate, mimic, or build on the epistemically and morally problematic features of eugenics past.

By contrast, and as our standpoint-driven discussion in chapters 3 and 4 intimated, the identification of certain sorts of people as bearing eugenic traits was part of a subhumanizing process. In terms that we introduced in chapters 5 and 6, the marked variation associated with eugenic traits was marked *for* segregation, marginalization, and even elimination; certain sorts of people were marked *as* subhuman, and treated as such. The lessons from the eugenic past for our thinking about contemporary practices of human enhancement are different here. For those sharing a standpoint eugenics, it is the discriminatory, subhumanizing, and eliminative dimensions to The Eugenic Mind that take center stage and that persist in continuing thinking about human enhancement. Rather than extracting the goals of human improvement, betterment, or enhancement from a eugenic past and looking to insulate their implementation from the excesses and mistakes of the past, proponents of a standpoint eugenics view the differentially targeted melioration within The Eugenic Mind itself as the core of the problem with eugenics past.

The idea of *newgenics*, a term sprinkled throughout the recent literature on eugenics, is often used to gesture at the relationship between the eugenic past and present practices. Indeed, I quietly injected that term into my discussion in each of the chapters in part I. Discussions of newgenics are focused largely on practices that deploy reproductive technologies and medical advances with differential impacts on the sorts of people who were the more overt target of past eugenic ideas, practices, and policies.

The appeal to newgenics, with its invocation of the eugenic past, minimally has the objective of arresting a kind of complacency about eugenics, namely, that it *is* primarily a matter of the past. Viewed in this way, discourse around newgenics looks chiefly to raise questions about what similarities and differences there are between eugenics past and present ideas, practices, and policies, and to advocate to ensure that present and future lives be safeguarded from the errors of that past, whether they be scientific or evaluative, epistemic or ontological, in nature. Newgenics then is something—like climate change or global free trade—that we can explore in the spirit of objective, neutral inquiry, a topic with many facets to it about which reasonable and rational people might disagree.

Yet for those who already feel the closeness of the eugenic past in their day-to-day lives, such as eugenics survivors, this perspective on newgenics is problematic. For them, the end of the explicit eugenic era was more a rebranding than an abandonment of eugenics. For them, eugenics went underground, with the attitudes and views that drove eugenics merely resurfacing in new guises as emerging technologies were sometimes chiefly motivated by and typically adapted for the eugenic purpose of eliminating certain sorts of people. Instead of being characterized as unfit or degenerate, they were regarded as less healthy or suffering from medical irregularities or abnormalities. Instead of state-mandated practices of euthanasia and sterilization being advocated with respect to them, there were practices of prenatal screening and selective abortion offered as matters of individual reproductive choice. From this standpoint, the term *newgenics* picks out eugenics achieved by other means. It is how The Eugenic Mind manifests itself in contemporary society.[1]

7.2 The Prenatal Back Door to Eugenics

Prenatal screening for genetic abnormalities, such as trisomy 21, and somatic or bodily abnormalities, such as spina bifida, have been regularly encouraged as a part of prenatal care within the medical and health professions for over twenty years; they are now a routine part of family planning and individual reproductive choice. The combination of prenatal screening followed by selective abortion of a fetus found to have genetic or somatic abnormalities has become a kind of paradigm case for the standpoint eugenic view of newgenics. Consider the disability studies scholar Marsha Saxton, a person who has spina bifida, on this combination: "The message at the heart of widespread selective abortion on the basis of prenatal diagnosis is the greatest insult: some of us are too flawed in our very DNA to exist; we are unworthy of being born. ... [F]ighting for this issue, our right and worthiness to be born, is the fundamental challenge to disability oppression; it underpins our most basic claim to justice and equality—we are indeed worthy of being born, worth the help and expense, and we know it!" Saxton's view here is typically taken to state "the expressivist objection" to selective abortion, and I will return to discuss that view later in the chapter. But first I want to provide more background to the practice, particularly as it operates in its best-known case, that of Down syndrome.

As we saw in chapter 4 in our discussion of the practice of human experimentation that formed a part of Alberta's eugenic past, Down syndrome is a nonhereditary, genetically based condition that is associated with male sterility and with reduced fertility in females. Down syndrome is the most common cause of intellectual disability, accounting for approximately 20 percent of all known instances. It is also known as trisomy 21 after the genetic condition that has been used through prenatal testing and screening to identify individuals with the syndrome: an extra copy of at least part of the twenty-first chromosome that is present in about 95 percent of all cases of Down syndrome. Children born with Down syndrome typically have mild to moderate intellectual disability, distinctive facial features, and face a higher risk of congenital heart defects and several other susceptibilities, such as to thyroid dysfunction and infection due to a compromised immune system. As with other children with trisomies, their life expectancy has increased dramatically over the course of the twentieth century due to both medical advances, such as antibiotics and open-heart surgery, and societal changes, with life expectancy shifting from nine to eleven years in the first quarter of that century to sixty years today. Quality of life measurements likewise show reported levels that approximate those of the general population, departing from earlier assessments of well-being.

The incidence of Down syndrome is approximately 1 in 800 births. This incidence rate can be put in perspective by considering it against other conditions regarded explicitly or implicitly as eugenic traits that were medically labeled at roughly the same time as Down syndrome. As table 7.1 indicates, Down syndrome is roughly an order of magnitude, that is, 10 times, less prevalent than cerebral palsy and roughly two orders of magnitude, 100 times, less prevalent than epilepsy.

The incidence of Down syndrome is unevenly distributed demographically. The best-known demographic pattern—the relationship between Down syndrome and maternal age—has been known for long enough for it to be one of the best-known cases of a fetal condition for which there is age-related screening and testing (e.g., amniocentesis). Although there is variation in the precise numbers that researchers have found, the following figures are representative of termination rates following the prenatal detection of trisomy 21: of all the Down syndrome cases detected prenatally, 88 percent of those in Europe and around 85 percent of those in the United States result in a termination of pregnancy.

Despite the high incidence of trisomy 21 in fetuses carried by pregnant women over age thirty-five, the majority of children born with Down

Table 7.1
Down syndrome compared to some eugenic traits

Condition	Discovery	Prevalence rates	Remarks
Cerebral palsy	William Little 1860	0.2%	Motor condition; associated with epilepsy 35% of the time
Down syndrome	John Langdon Down 1862, 1866	0.02%	"mongolism"; genetic specification as a trisomy in 1959; the most widely tested-for condition in prenatal screening; nonheritable
Epilepsy	also called "the sacred disease"; 19th-century medical articulations	3%	Defined chiefly in terms of presence of seizures, i.e., behaviorally; appears in Pennsylvania (1905) and Virginia (1924) sexual sterilization laws
Huntington's disease	Huntington 1872; earlier work 1840s	0.05%	Autosomal dominant, late onset; identification at chromosome 4 in 1983; proposed ground for eugenic sterilization by Davenport (1915)

syndrome have mothers who are younger than thirty-five, since the incidence of pregnancy itself is significantly higher in that younger group. Put differently, although the incidence of trisomy 21 rises with maternal age, the incidence of pregnancy falls with maternal age sufficiently that prenatal screening practices that focus on older mothers do not detect the majority of fetuses that have trisomy 21.

In 2007, the Society of Obstetricians and Gynaecologists of Canada (SOGC) issued a clinical practice guideline that recommended that the existing reliance on maternal age as a minimum standard for prenatal screening was inadequate and should be removed as an indication for more invasive testing, such as amniocentesis, which carries with it a direct risk to the fetus. In place of this reliance on maternal ages, the guideline's principal recommended outcome is to "offer non-invasive screening for Down syndrome or trisomy 18 to all pregnant women"; that recommendation is retained in the 2011 and 2012 updates to this guideline. While such practice guidelines are not binding, either legally or morally, in terms of screening practices in Canada, it both reflects and influences those practices.

SOGC clinical practice guidelines are constructed carefully by a distinguished set of authors in consultation with larger committees on genetics, prenatal diagnosis, and diagnostic imaging and with the review and approval of the executive and council of the Society of Obstetricians and Gynaecologists of Canada, and by the board of the Canadian College of Medical Geneticists. The 2007 guideline includes a review of existing screening practices in Canada, drawing on papers published in the preceding twenty-five years in medical and related professional journals, as well as introductory explanations of key notions, such as what screening is and its relationship to individual diagnosis, and reports on screening detection and false positive rates. The abstract states that the guideline is "intended to reduce the number of amniocenteses done when maternal age is the only indication" having the benefit of "reducing the numbers of normal pregnancies lost because of complications of invasive procedures." For all that attention to detail, most striking are two points of inattention.

The first point is that nowhere in this practice guideline is any space devoted to describing what Down syndrome *is like*. By this I mean not only a characterization of what it is like for an individual or a family to "live with Down syndrome," but also even a description of the medical symptoms of, or variation in the symptoms of, those individuals with Down syndrome. One might think that there is no real need to provide such descriptions in a practice guideline. Yet what the absence of such detail conveys, at least implicitly, is the view that the only thing that at least obstetricians and gynecologists need to know about infants, children, and adults who have been diagnosed as having Down syndrome is that they are infants, children, and adults with Down syndrome. What they already share as common knowledge with members of the general public is that a diagnosis of Down syndrome is serious enough to warrant testing for its presence prenatally with an expectation of termination.

The second point is that despite the explicit emphasis on the goal of reducing the number of "normal pregnancies lost," the most obvious effect of the recommendations remains implicit: to reduce the number of babies born with Down syndrome, and thus the number of infants, children, and adults who "live with Down syndrome." Together with the first point, this provides a basis for viewing the combined practice of prenatal screening for trisomy 21 together with the expectation of termination should the screen detect a fetus with trisomy 21 as what the well-known former U.S. Surgeon General C. Everett Koop and the conservative political commentator George Will popularized as "search and destroy" missions, exemplifying a kind of eliminative attitude toward Down syndrome as making for life not worthy of life.

Technologies such as preimplantation genetic diagnosis (PGD) and in vitro fertilization (IVF) are capable of screening for disability while avoiding the more pointed moral quandaries of selective abortion. PGD is already being used to profile embryos with Down syndrome and "neural tube defects," such as spina bifida; a newgenic future where prospective parents can select from a set of embryos the most desirable (or "enhanced") before implantation is not that far away.

The bioethicist and philosopher Robert Sparrow has argued that an even more extensive and deliberate form of newgenics, which he terms "*in vitro eugenics*," is lurking on the biomedical horizon. Scientists hypothesize that in the not-too-distant future it will be possible to create human gametes from human stem cells. Sparrow argues that this technology will enable the in vitro production of multiple generations by repeatedly deriving and then combining gametes from two sources: the stem cells of the newly formed embryo, and different stem-cell lines. Proceeding through multiple generations of embryos in the laboratory before implantation holds the possibility of producing desired genotypes and deliberately raising the quality of "human stock" without the messy business of nonconsensual sterilization and selective abortions. Moreover, unlike PGD, and as Sparrow points out, in vitro eugenics is not limited by the chance recombination of genes, but is a deliberate selection process.[2]

7.3 Eugenic Subhumanization and a Continuing Preoccupation

The continuing preoccupation with the science of genetics, and with the control and direction of human populations through an enhanced grasp of human genetics, has been modified in light of shifts in the science of genetics itself, particularly as it has progressed beyond Mendelian genetics to incorporate population and molecular models of genetics. This progression initiated a shift among eugenicists away from the characterological study of types of people that marked early forms of eugenics toward a statistical consideration of human traits triggered by specific genetic markers. As eugenicist Frederick Osborn explained already in 1940, "Eugenics in a democracy seeks not to breed men to a single type, but to raise the average level of human variations, reducing variations tending toward poor health, low intelligence, and anti-social character, and increasing variations at the highest levels of activity." Osborn was president of the American Eugenics Society, an organization that went on to change its name to the Society for the Study of Social Biology in 1972, having already changed the name of its journal from *Eugenics Quarterly* to *Social Biology* in 1969, the same year that Britain's *Eugenics Review* changed its name to the *Journal of Biological Science*. (It took the British Eugenics Society until 1988 to make the corresponding organizational name change, when it became the Galton Institute.)

A focus on statistical traits at the level of populations recognizes the genetic diversity within groups, and thus purportedly steers clear of the racist, classist, and ableist typology of early eugenics. At the same time, molecular genetics locates desirable and undesirable human traits at the level of DNA sequences. As its director James Watson stated, the Human Genome Project emboldened the search for genetic markers of disability in the hope of "banishing genetic disability." Going molecular, like focusing on statistical traits, allows for the pursuit of the general eugenic goal of improving human populations while ostensibly avoiding the categorization of *types* of "less desirable" people tracked through blood lines. When all is settled, the fact remains that intellectually disabled people are disproportionally targeted by newgenic practices. Cognitive disability may no longer be a sub-human *kind* in the scientific and bioethics literature as feeble-mindedness was, but it remains an especially undesirable trait.

Another way of understanding the transition from eugenics to newgen-ics is through what Rosemarie Garland-Thomson has termed *eugenic logic*— the belief that "our world would be a better place if disability could be eliminated." Perversely echoing Hannah Arendt's indictment of Adolf Eich-mann in *Eichmann in Jerusalem* for claiming "any right to determine who should and should not inhabit the world," eugenic logic asks: "Why should the world we make and occupy together include disability at all?"

Garland-Thomson argues that while eugenic logic manifests itself in a wide array of practices and discourses—from segregation to extermination and from practical health programs to social justice initiatives—at its core it is necessarily eliminativist. That is, no matter the means of execution, disability both can and should be rooted out of populations to produce a better world. The utopian edge to eugenic logic follows from the ren-dering of disability as an inevitable suffering and a tragic mutation of the human condition. Disability in this reading is subhumanizing, alienat-ing us through pain, stigma, suffering, dependency, and limitations from our status as proper humans. As the disability studies scholars David Mitchell and Laura Snyder have suggested, disability is the master trope of human disqualification. The ubiquitous and presumptive nature of eugenic logic—for example, *obviously*, high IQs are better than low IQs; *obviously*, being sighted is better than being blind—provides a basis for skepticism about the prospects of overcoming the horizon of ableism through more disability-sensitive uses of advances in genetic knowledge and reproductive technology.[3]

7.4   Recasting Debate over the Expressivist Objection

Garland-Thomson's idea of a eugenic logic and its relationship to subhu-manization also sheds light on the ongoing debate over what is usually called the *expressivist objection* to prenatal screening followed by selective abortion. This objection arose in the 1980s as a central part of the disability rights critique of prenatal testing and selective abortion through the work of disability studies scholars, such as Adrienne Asch and Marsha Saxton. Asch and Saxton, among others, took the widespread and growing practice of selective abortion targeting fetuses with indications of later-life impairments to express a subhumanizing and damaging view of people with those impairments and disabilities. As we have seen, Saxton articulated this message as saying that "some of us are 'too flawed' in our very DNA to exist; we are unworthy of being born," being based on "the assumption that any child with a disability would necessarily be a burden to the family and to society, and therefore would be better off not being born." Since the objection was specifically to the practice of *selective* abortion and many of the most prominent advocates of the objection were in general supportive of reproductive autonomy, the development of the expressivist objection did not presuppose that the fetus had a right to life and reflected, rather, one point at which reproductive and disability rights clashed.

One way to represent the objection is as an explicit argument that begins with the following three premises: the first making a claim about the chief function of prenatal testing, the second specifying the expectations embedded in that practice, and the third drawing out one implication:

1. The practice of prenatal testing functions chiefly to detect fetuses that have a biological profile predictive of postnatal impairment.
2. The expectation (but not requirement) in individual instances of this practice is that a fetus with such a profile will be terminated, rather than carried to term.
3. Terminating a fetus primarily on the basis of its possessing a biological profile predictive of postnatal impairment implies the judgment that such a fetus is not worth carrying to term to become, in turn, a baby, infant, child, then adult with that impairment.

The word "expressivism" comes from two conclusions drawn from these premises, one concerning what the practice itself implies, the other concerning those who participate in the practice. To articulate these conclusions in some of their strongest versions, which reflect Saxton's articulation of the objection as previously discussed, consider (4) and (5) below:

4. Thus, the practice of prenatal testing so applied, and with that expectation, expresses the view that people with those postnatal impairments are not worthy of life.

5. Therefore, those who participate in that practice also express the view that people with those postnatal impairments are not worthy of life.

The natural and standard reply to such strong forms of the expressivist objection is to challenge either or both inferences to these conclusions. This has been done typically by identifying other practices (such as taking folic acid) that aim to prevent the birth of a child with an impairment (such as a spina bifida), but that are not taken to express either of these extremely negative views of people with those impairments. Acting to prevent the birth of a child with spina bifida by taking folic acid during pregnancy does not imply that people with spina bifida are not worthy of life. More generally, collective practices or individual actions that aim to prevent an undesirable outcome for a specific kind of individual need not express this strongly negative view of those individuals. This is true even when those practices and actions prevent such outcomes by changing the traits of those individuals that make them of that kind.

The subhumanizing connection between the practice of selective abortion on the basis of particular impairments and negative views of people with those impairments that the expressivist objection attempts to make can be preserved, however, in weaker expressions of (4) and (5) that still appeal to the eliminativist thinking at the heart of eugenic logic. Consider increasingly weakened forms of the negative attitude that (4) and (5) articulate, using just (4) illustratively:

4. Thus, the practice of prenatal testing so applied, and with that expectation, expresses the view that

    (a) people with those postnatal impairments are not worthy of life.

    (b) people with those postnatal impairments have a trait so negatively valued that its presence provides a sufficient reason to abort an otherwise desired pregnancy.

    (c) it would be better for people to exist without those postnatal impairments.

    (d) it would be better for those postnatal impairments not to exist.

    (e) those later impairments are strongly negative traits of the people who have them.

While the weakening from (4a) to (4b) shifts from talking of the life worthiness of people to the negative valence of the traits they have, it preserves the connection between selective abortion and negative, subhumanizing views of people with the corresponding traits by indicating the extent of the devaluation of those traits in eliminativist terms. It is not simply that traits such as having spina bifida are not regarded as neutral—as indicated by the unproblematic nature of the preventative practice of taking folic acid—but that they are sufficiently negatively valued to be difference-makers in decisions about pregnancy and termination. If we understand such a specification of the strength of the negative evaluation of such traits to be implicit in the further weakened conclusion (4c) to (4d), we can see a kind of eliminativist logic surviving in more modest versions of the expressivist objection. This eliminativist view marks a difference between preventative practices regarded as relatively morally unproblematic and the more contentious practice of selective abortion. While both reflect the view that it is better to be able-bodied than to be disabled—better to be disability-free than disability burdened—only the practice of deselection expresses the severity of the negative attitude about impairment and disability through its eliminativism. And that eliminativism provides a strong reason to view the practice of selective abortion as eugenics by other means, as eugenics underground.[4]

7.5   Outing Eugenic Logic in Bioethics: Agar and Savulescu

The contemporary bioethics literature readily makes transitions from the view that disability is not a neutral but a negatively valenced trait to the idea that we should eliminate disability where we can. In fact, we saw this kind of shift in chapter 4 where we briefly considered the views of the bioethicist and philosopher Julian Savulescu. Savulescu begins his discussion of parental rights and obligations with general considerations of what he calls "opportunity restriction" and providing one's children with the best lives they can have; he concludes, in part, by equating selecting the best children with selecting "those with the least [sic] disabilities" (2008, 66). I want now to return more specifically to critically discuss Savulescu's Principle of Procreative Beneficence, briefly introduced in that earlier discussion, and the basis it has for more substantial conclusions about disability and reproductive autonomy in Savulescu's thought and work.

Savulescu's discussion here forms part of the resurgence of sympathetic interest in eugenics within the philosophical bioethics community, an interest manifest in Philip Kitcher's "utopian eugenics," Rob Sparrow's "in vitro eugenics," and Nicholas Agar's "liberal eugenics." As we saw foreshadowed in section 7.3 by Frederick Osborn's appeal to "eugenics in a democracy," with its emphasis on higher levels of variation rather than a preferred type of people, such advocates of newgenics attempt to sever the old from the new, aiming to foster human improvement without explicitly devaluing certain types of human lives.

Consider Agar's liberal eugenics. In embracing a radical form of reproductive autonomy, proponents of liberal eugenics spurn state involvement in the figurations of a good human life and advance "the development of a wide range of technologies of enhancement ensuring that prospective parents were fully informed about what kinds of people these technologies would make." In perfect pluralistic, liberal fashion, parents' *individual* conceptions of the good life should direct the selection of preferable traits for their child, and parents "will acknowledge the right of their fellow citizens to make completely different eugenic choices."

This laissez-faire approach to eugenics has struck some, however, as incapable of avoiding both the horrors of early twentieth-century eugenic practices and the problematic aspects of selective abortion discussed in section 7.4. Indeed, in more recent work Agar himself seems to endorse the kind of eliminativist thinking we identified there. After noting that the use of prenatal genetic diagnosis "enables parents to make a narrow range of choices about the genetic constitution of their future children," Agar says: "Suppose that the technology is mainly used to avoid passing on genetic variants linked with serious diseases. Individual eugenic choices will then have the effect of improving human stock. We may therefore achieve Galton's desired outcome without selective breeding."

Those individual eugenic choices—to choose "healthy" over "diseased" embryos for implantation—do not so much avoid selective breeding, however, as locate it earlier in the chain of reproductive choice. Rather than killing defective infants through euthanasia, as we saw Harry Haiselden advocate one hundred years ago in American medicine, or selecting against fetuses that are diagnosed as having a genetic profile indicative of later impairment, here the form of negative eugenics in play is directed at embryos with a corresponding biochemical signature. In all three cases, individual eugenic choices do not simply collectively cause a eugenic outcome, but are themselves the means through which eugenics operates in a neoliberal marketplace.

Telling here is Agar's presumption that individual choices "to avoid passing on genetic variants linked to serious diseases" would *thereby* contribute to the collective goal of "improving human stock." Here Agar assumes an antithetical relationship between having *any* such variant and the meliorative project of human improvement. The sweep of this generalization should at least serve as a flag for those wary of the translation of the negative valuation of a trait into eliminativist practices. While the eradication of some such diseases, such as Tay-Sachs disease and other early-onset, life-shortening diseases and conditions, unquestionably contributes to that meliorative project, the overwhelming majority of "genetic variants linked to serious diseases" stands in a more complicated and problematic relationship to the goal of human improvement. In part this is because the linkages between genetic variation and serious disease are pervasive, often probabilistic, and mediated by genetic, somatic, and external environmental factors. But it is also because even "serious diseases" with clear, localized decisive genetic causes themselves range from the extreme case of Tay-Sachs disease, where eradication seems relatively uncontroversial, to those that are compatible with lives that are healthy and fulfilling until the disease's onset in mid- to late-life (Huntington's disease being one well-known example). Hereditary blindness, deafness, and epilepsy—to take eugenic traits that would typically be regarded as serious, hereditarily transmissible diseases and that featured in the central Nazi sterilization law—are three further examples that should give pause to whether Agar's "individual eugenic choices" will improve human stock in ways that are more acceptable than state-mandated eugenics. Given ableism, individual decision makers under the influence of eugenic logic will likely make choices that equate disability with suffering and pity, and as something to be avoided or eliminated at even relatively high cost.

Savulescu takes a different approach to Agar, arguing that autonomy is simply not a strong enough principle to guide reproductive ethics. For Savulescu, liberalism offers little bulwark against morally implausible views of the good life. As we saw in chapter 4, central to his view of reproductive ethics is the Principle of Procreative Beneficence, which in joint work with the bioethicist Guy Kahane has been expressed as follows: "If couples (or single reproducers) have decided to have a child, and selection is possible, then they have a significant moral reason to select the child, of the possible children they could have, whose life can be expected, in light of the relevant available information, to go best or at least not worse than any of the others" (274). This principle claims that given a choice, parents have a moral obligation to produce children with the best chance of the best life. Savulescu believes that the principle should guide parental reliance on prenatal genetic diagnosis and selective abortion in the case of many commonly recognized disabilities.

The Principle of Procreative Beneficence has been challenged in various ways, ranging from appeals to acceptable thresholds of well-being and the rationality of random selection, to the expressivist critique of selective abortion that we have just considered. Rather than focus on whether Procreative Beneficence itself is true, here I want to show that those who *apply* the principle can also be challenged on their own terms. So let us provisionally grant Procreative Beneficence and focus instead on clarifying and then challenging reasoning that begins with that principle and then draws a conclusion about disability, well-being, and choosing children.

Like Agar's conception of a liberal eugenics, Savulescu's Procreative Beneficence is stated in disability-neutral language, and like Agar, Savulescu is very much aware of the sorts of choices that he expects to be made, assuming Procreative Beneficence. Savulescu and Kahane believe that what is best for a child, what will increase that child's well-being, is to be created disability-free. They say, of "paradigmatic cases of disability in the everyday sense—deafness, blindness, and intellectual subnormality" (288–289), that they are "inclined to believe most of these are disabilities [in their sense] in the conditions holding at present and in the foreseeable future" (289), where they idiosyncratically define disability, in part, as a property that "leads to a significant reduction in [a subject's] well-being in circumstances C, when contrasted with realistic alternatives" (286). Thus, if parental choices are governed by Procreative Beneficence, they claim, they will have a significant moral reason to select the child who is disability-free, over one who is not.

The crucial claim that disability reduces well-being admits of a weaker and a stronger reading. Each reading has a different implication for how to reconstruct the argument from Procreative Beneficence to the claim about disability. On the strong reading, the claim says that disability reduces well-being *substantially*; on the weak reading, it says only that disability reduces well-being *to some degree or other*, allowing that such reductions might be relatively small. We can be more explicit here. The conclusion about disability that Savulescu and Kahane argue for is what I have elsewhere called *Disability-Free Procreation*.

Disability-Free Procreation: Prospective parents typically have a significant moral reason to select against creating a child who will probably develop a disability, provided they could instead create a child who probably will not develop that disability.

Table 7.2 below summarizes two versions of the argument from Procreative Beneficence to Disability-Free Procreation, one that draws on the stronger version of the claim about disability and the reduction of well-being (Version A), the other drawing on the weaker version of that claim (Version B).

One way to think of the difference between these two versions of the argument is to consider which of their claims are most ambitious. In Version A, the more ambitious claim is the second premise, whose strong claim about disability and reductions in well-being is difficult to defend.

Table 7.2
Two versions of the argument for disability-free procreation

Argument for Beneficence in Action Version A	Argument for Beneficence in Action Version B
A1. If most disabilities reduce well-being *substantially*, then prospective parents typically have a significant moral reason to select against creating a child who will probably develop a disability, provided they could instead create a child who probably will not develop that disability. A2. Most disabilities reduce well-being *substantially*.	B1. If most disabilities reduce well-being *to any degree*, then prospective parents typically have a significant moral reason to select against creating a child who will probably develop a disability, provided they could instead create a child who probably will not develop that disability. B2. Most disabilities reduce well-being *to some degree*.
Disability-Free Procreation: Prospective parents typically have a significant moral reason to select against creating a child who will probably develop a disability, provided they could instead create a child who probably will not develop that disability.	*Disability-Free Procreation*: Prospective parents typically have a significant moral reason to select against creating a child who will probably develop a disability, provided they could instead create a child who probably will not develop that disability.

By contrast, in Version B, this strong claim is weakened, with the more ambitious premise being the first premise, which is a strong form of the Principle of Procreative Beneficence. But that version of the principle is itself implausible, given that small reductions in well-being do not, in general, generate moral reasons for parental choice of the kind envisaged. Properly spelled out, these arguments that appeal to Procreative Beneficence to reach conclusions about parenting and disability falter. They do so in part because they shift between or blend these two versions of the argument for Disability-Free Procreation.

While Savulescu and Kahane concede that the constitution of a "best life" upon which Procreative Beneficence is premised remains highly contestable and cannot be agreed upon positively, they insist that "there is considerable consensus about the particular traits or states that make life better or worse, a consensus that would rule out many procreative choices as grossly unreasonable" (279). This "considerable consensus" operates within the milieu of eugenic logic that construes disability as a tragic suffering.

Moreover, Savulescu and Kahane's reliance on eugenic logic leads them to trite conclusions such as: "We all vary in our abilities and our disabilities. To a degree, we all suffer from disability." Such sentiments erase from our consciousness the lived realities of disability, often situated along multiple axes of oppression, and paradoxically usher in its eradication. Stating that "we all suffer from disability" veils the eliminative abortion rates of Down syndrome fetuses and the fact that autistic self-advocates are currently fighting for their survival against well-funded organizations such as Autism Speaks that are, by contrast, actively working to "cure" and ultimately stamp out autism (see section 7.8).

Savulescu and Kahane echo Agar in erecting a public/private divide to distinguish newgenic practices from the unfortunate and misguided eugenic practices of the early twentieth century. Procreative Beneficence, like liberal eugenics for Agar, steers clear of eugenics—a term Savulescu refuses to employ in relation to his principle—insofar as eugenics has been characterized by Savulescu elsewhere as a state-level project of producing a better population through coercive tactics. Reproductive choices and Procreative Beneficence are rather a distinctly *private* enterprise that aim not at producing better populations, but a best child through morally informed choices of what constitutes a good life. The aggregate effects of individual choices informed by eugenic logic do not register for Savulescu as morally relevant. As a private enterprise, reproductive choices are not political considerations.

However, neither Savulescu and Kahane nor Agar appreciate that eugenic logic is agnostic toward social organization: it is highly adaptable, working just as well bottom-up as top-down. In early twentieth-century discourse and practices, eugenic logic circulated explicitly and unabashed. It has since gone underground and, in Troy Duster's terms, snuck in through the back door. In our liberal, postcolonial, and post-Holocaust society we now talk benignly of human lives that are "advantaged" or "preferable" rather than seeking human perfection measured against a hegemonically normal subject. Yet the underlying eliminativist logic remains largely the same: the world would be better not simply without eugenic traits, but also without people with disabilities.[5]

7.6   Diversity and Neoliberalism

The Eugenic Mind might have become a purely historical legacy left behind as the ideological and socioeconomic shifts that occurred after the heyday of eugenics in the early twentieth century—from the welfare state to neoliberalism, and from a disciplinary to a control society—shaped up new social structures and institutions. But I have been arguing, by contrast, that The Eugenic Mind itself has had a continued existence, filling new social molds. In this section and the next I want to suggest that far from being undermined or made otiose by neoliberalism and a control society, The Eugenic Mind has altered little, finding augmentation and depoliticization in the ideology of neoliberalism.

As indicated in chapters 5 and 6, normalcy has been a central concept used to sort those deemed worthy of inhabiting the world from those who are not, thanks in part to the work of Lennard Davis, particularly his *Enforcing Normalcy*. Yet while normalcy is still often quoted as driving contemporary eugenic practices, this cannot simply be assumed. Davis himself has more recently argued that diversity is replacing normalcy as the reigning ideology in the cultural imaginary. The shift away from a universal standard of embodiment toward a liberal valuing of heterogeneity is captured well, Davis notes early in his more recent work, by a recent slogan for Tylenol: "get back to normal ... whatever your normal is.™" Within a postcolonial and globalized society, the concept of normalcy has slipped from being a hegemonic gold standard in the public realm. More important, and as this advertisement hints at, "normalcy" is too monolithic a category for neoliberalism to exploit optimally. The ideology of diversity, on the other hand, generates an endless configuration of marketable traits through which we become increasingly enmeshed within neoliberalism. Trading on normalcy is no longer profitable, so human variation has—quite literally in this example—been trademarked. In a final move that reanimates a universal human identity, neoliberalism reduces diversity to consumption through the logic of market functioning. As neoliberalism would have it, race, gender, and disability dissolve under the common banner of consumption.

Although Davis is not explicitly concerned with eugenics in this more recent analysis, I want to suggest that the (ostensible) displacement of normalcy provides an important cue for understanding contemporary eugenic practices. Consider first that for Davis, the sustainability of diversity as an ideology requires the suppression of the abject or hyper-marginalized. As he argues, "You can't have a statement like 'we are different, and we celebrate that diversity' without having some suppressed idea of a norm that defines difference in the first place." "Diversity" must quell the spectacularly and incurably disabled in order to maintain itself. Davis points out that it is difficult to imagine neoliberalism celebrating "homeless people, impoverished people, end-stage cancer patients, the comatose, heroin, crack, or methamphetamine addicts." These, and other custodial forms of disability such as Tay-Sachs and spina bifida, are resistant to the flow of neoliberal capital; they are the "outside" that must be excluded to make "diversity" culturally and economically salient.

Put otherwise, the hyper-marginalized are the remainders of a system—what Ehrenreich and Ehrenreich long ago termed the medical-industrial complex—in the business of investing highly mobile and malleable subjects with capital. The rapid growth and consolidation of corporatized medicine has led to attenuated rather than increased care. This is to be expected from an industry driven by profit weighed only against the need for "healthy" and productive human capital required for the stability of private capital, as Estes, Harrington, and Pellow have argued. The totalizing logic of neoliberalism thus marks the fantastically and incurably disabled as disposable, or, as Davis suggests in reference to Giorgio Agamben, as *zoê*, bare life. It is no surprise that the choice *not* to select against Tay-Sachs or spina bifida is unintelligible within this milieu.[6]

7.7 Eugenics as Private Enterprise

By itself, the hyper-marginalization of fantastically and incurably disabled bodies results in nothing new, a rehashing of eugenic anxieties for a new era. However, the exclusion of these bodies makes space for an emergent form of eugenics under the banner of difference. Despite Davis's recent claim that the normal is no longer the structuring trope of disability, and that disability itself is too rigid a social category ever to be included within the neoliberal category of "diversity," disability lives on under neoliberalism. Jasbir Puar has argued that although the binaries of normal/abnormal, abled/disabled have been increasingly dismantled by neoliberal formulations of health, agency, and choice, all bodies "are being evaluated in relation to their success or failure in terms of health, wealth, progressive productivity, upward mobility, enhanced capacity" and thus there is "no such thing as an 'adequately abled' body anymore." The gradation and constant modulation of "ability" is profitable for capitalism insofar as it is not attainable once and for all. Rather, in "neoliberal, biomedical, and biotechnological terms, the body is always debilitated in relation to its ever-expanding potentiality." While Davis's claim is incomplete, Puar's formulation of "debility" as a continual and productive modulation of bodily capacities conversely seems to turn on Davis's recognition of the erasure of hyper-marginalized bodies.

From this perspective, Savulescu is not wrong to conclude that "to a degree, we all suffer from disability." The first segment of this statement, "to a degree," acknowledges the necessary exclusion of certain forms of disability from the neoliberal equation, combining with the following universalization to gesture at the shift, as foreshadowed by Donna Haraway, from perfection to optimization. Within this logic, biomedical enhancements such as genetic therapy do not aim at normalcy but, as Savulescu indicates throughout his discussion of Procreative Beneficence, at fluid and infinitely productive notions of health, enhancement, and advantage. These notions and corresponding bodies are deeply invested with neoliberal capital, sustaining entire industries at often tremendous economic cost to individuals that further stratifies socioeconomic inequalities.

We have seen that Agar and Savulescu distance their reproductive ethics from eugenics by emphasizing that eugenics offered public-interest justifications for intervening in reproduction, while Procreative Beneficence, and by extension, liberal eugenics, are explicitly *private enterprises*. The irony of this claim is that since the era of Reagan and Thatcher, direct state provision has progressively given way to the privatization of healthcare and biomedicine. Similarly, Savulescu maintains that "it was the eugenics movement itself which sought to influence reproduction, through involuntary sterilization, to promote social goods." Yet Savulescu fails to appreciate the increased outsourcing of social goods to the private sector, and more specifically, the outsourcing of biosocial goods (or in Foucauldian parlance, biopower), to the medical-industrial complex. Eugenic practices and logic have not gone away when one considers the sustained attention given to genetic counseling and human augmentation as major opportunities for investment and profit growth. Agar and Savulescu have just not been watching the back door.

It is therefore important to recognize that under the watchful (or perhaps neglectful) eye of neoliberalism, newgenics has become dangerously depoliticized. Recall that Agar places the onus on those who would restrict reproductive choices to prove harm, while Savulescu and Kahane argue that it is "up to us whether we love our children and give all people in society a fair go. This need not be affected by decisions about selecting which people come into existence." Not only is disability justice here divorced from questions of reproductive ethics and who should and should not inherit the world, but also the latter is rendered an individual (market) choice rather than an issue of justice at all. The hyper-reliance upon the self rather than community or the state, congruent with the neoliberal flattening of power relations, is what Judith Butler has referred to as "responsibilization," and what Angela Davis has recently described as a "neoliberal conspiracy designed to keep us from connecting with one another and bringing about change." In this regard, my earlier comment that eugenic logic is agnostic regarding social organization should be amended. Sneaking through the back door, eugenic logic thrives when its manifestations are finely distributed throughout the social field and reduced to individual, private choices. It is possible that the eliminativism at the heart of eugenic logic works even better bottom-up than top-down.[7]

7.8   Eugenic Techniques of Silencing

From the perspective afforded by standpoint eugenics, the reigning cautious optimism about our collective ability to avoid problematic eugenic consequences in our increasingly privatized reliance on reproductive biotechnologies transgresses the boundary between confidence and complacency. People with certain traits—the advance signs of which are sufficient to warrant uncritical and routinized practices of termination manifesting a kind of eliminativist mindset—readily view themselves as survivors of eugenics, not so much in past but in ongoing forms. For them, The Eugenic Mind lives on, and Garland-Thompson's eugenic logic highlights the stakes for disability politics. While we must continue to fight for nonnegotiable issues of accessibility and inclusion as full citizens, it is important to be mindful that our survival, our right to inherit the world, is increasingly at risk. Countering that complacency involves a wider net of critical attitudes that recasts even technologies such as cochlear implants and prenatal genetic diagnosis from being benign, restorative technologies to functioning as tactical technologies that aim to erase disability from our consciousness and ultimately the human condition.

Reading disability politics through the history and future of eugenics thus requires that we pursue operative sites of eugenic logic within which it can be contested. A central, yet unstable, tactic through which eugenic logic functions is suppressing any voices that deny that disability is tragic suffering to be avoided at all costs, and that question or fail to see the distinctive value of survivor knowledge in our understanding of eugenics past and present (see chapter 9).

Consider also the experience of autistic self-advocates. Organizations such as the Autistic Self-Advocacy Network (ASAN), alongside a host of autistic self-advocate bloggers—Autistic Chick, Aspie Rhetor, Tiny Grace Notes (Ask an Autistic), Neuro Queer—relentlessly criticize Autism Speaks, a major international nonprofit organization that sponsors autism research. While Autism Speaks putatively advocates for autistic individuals and their families, neurodiversity community members unanimously condemn the nonprofit group for fear-mongering media projects such as "I Am Autism" and "Autism Everyday," for funneling large percentages of its income into prevention technologies, and for dehumanizing rather than representing the lives of autistic people. Autism Speaks, they argue, comprehends autism merely as a problem to be solved.

A recent editorial by Suzanne Wright, the founder of Autism Speaks, demonstrates the eugenic logic of this organization that many autistic individuals find so deplorable: "This is autism. Life is lived moment-to-moment. In anticipation of the child's next move. In despair. In fear of the future." Wright desperately calls for a national plan—"for the children" and "for the future"—to effectively rid the world of autism.

Despite their sustained and well-supported critiques, self-advocates are silenced. Autism Speaks has never explicitly acknowledged the voices of neurodiversity self-advocates, even though the organization has at times plagiarized their work, as Alyssa Hillary has pointed out. Moreover, organizations such as Autism Speaks discredit self-advocates by creating a binary between "low-functioning" and "high-functioning" autistics, where the former "severe" and often nonverbal autistics represent what autism "truly" is, while the latter can neither possibly understand nor represent autism. Not only is the binary itself decried by self-advocates, but as Autistic Chick has noted, there are *many* so-called "low-functioning" self-advocates who have taught themselves to communicate and strongly resist Autism Speaks.

The silencing of disabled people under eugenic logic is accordingly two-pronged. On the one hand, the *sites* of knowledge production are discursively controlled. Oral histories and blogs are rendered unreliable and unauthoritative, lacking peer-review processes and the corroboration of data by "experts" who provide a discursive stamp of approval. On the other hand, the authority of the voices of those targeted by eugenic logic is continually undermined and discredited. Neoliberal organizations such as Autism Speaks add an economic valence to this equation. That is, discrediting the voices and sites of knowledge production of autistics is necessary to suppress any opposition to the big business of the medical-industrial complex.[8]

7.9   Conserving Disability

The response to eugenic logic that Garland-Thomson herself has made is to advocate actively for conserving disability—for identifying disability not as a disadvantage or liability but as a resource that must be encouraged to flourish. Contrary to what eugenic logic contends, disability is neither a subnormal individual form that humanity can take nor an extraneous facet of the human community that can or should be eliminated in a utopian future. Disability is instead a fundamental aspect of humanity, one to be conserved and even encouraged. As Garland-Thompson explains, "The idea of preserving intact, keeping alive, and even encouraging to flourish denoted by *conserve* suggests that the characteristics, the ways of being in the world, that we think of as disabilities would under such a definition be understood as benefits rather than deficits" (italics in original). Likewise, the disability studies scholar Alison Kafer has argued that "to eliminate disability is to eliminate the possibility of discovering alternative ways of being in the world, to foreclose the possibility of recognizing and valuing our interdependence." If disability is a generative and intrinsic aspect of the human condition, there is much to lose if disability and disabled people are eliminated from the world.

Garland-Thomson organizes the transformative potential of disability under three rubrics: disability as narrative resource; disability as epistemic resource; and disability as ethical resource. The first appeals to disability as a common belonging that both unsettles and transforms pejorative cultural narratives of disability. The second speaks to the privileged and unique forms of knowledge production that follow from disabled subjects' particular, though threatened, mode of embodied being-in-the-world. The third encourages openness to the contingency and lack of expectancy represented by disability as a model for human flourishing.

While Garland-Thomson does not explicitly take up the issue of silencing, it is never far away in her insistence that disability be mobilized as a narrative and epistemic resource. However, I suggest that the appeal to disability as an epistemic resource be expanded upon to articulate more clearly the critical task of contesting and reclaiming sites of knowledge production. As we'll see in more detail in chapters 9 and 10, standpoint epistemology provides a critical—even if in some respects a problematic—conceptualization of disability knowledge. Starting with the assumption that *all* knowledge is situated, standpoint epistemology positions itself against dominant philosophies of objectivity and epistemic neutrality. From this starting point, standpoint epistemology inverts power structures such as sexism, classism, racism, or ableism by asserting that that those marginalized by systems of oppression have better insight into how the system works than those who benefit from its operation. In this case, standpoint epistemology resituates eugenic survivors and autistic self-advocates as possessing privileged epistemic resources vis-à-vis manifestations of eugenic logic.

If disability politics is to conserve disability in the face of ongoing eugenic practices and a eugenic logic that seeks to "trade the present in on the future," we must continue to make recourse to a past that ripples through to the future. A key way of conserving disability in light of the marginalization of voices from the eugenic past is to take seriously the central message of standpoint epistemology. Those who have lived that eugenic past, together with their successors whose lives have been most directly shaped by the legacy of eugenics have much to teach us all about eugenics and disability and their relationship in The Eugenic Mind.[9]

8 Eugenics as Wrongful Accusation

8.1 Persistent Eugenic Pasts

I began this exploration of the persistence of eugenics in chapter 5 by articulating a puzzle—the puzzle of marked variation—and by suggesting that the limits of the appeal to biopolitics as a response to this puzzle cast a shadow over popular social constructivist views of eugenics and disability. I then turned, in chapter 6, to introduce and then defend what I called a socio-cognitive framework for addressing that puzzle, one that draws on, and draws out, both social and psychological dimensions to our views of marked variation. In chapter 7, I sought both to convey why adopting a standpoint eugenics perspective implies a tight relationship between the eugenic past and the newgenic present, and to show the persistence of eugenic ideas in contemporary practices, including in dominant discourses within bioethics and philosophy.

In completing this exploration of eugenics' persistence, this chapter follows the general trajectory already established, even recapitulating something like the major activities undertaken in all three preceding chapters, albeit with a different subject matter. Here I begin with a sort of puzzle— or at least a puzzling question—about our eugenic pasts, not so much *where do ideas of human variation come from?* as *how is it that eugenic practices didn't just disappear or cease?* In offering an answer to this question, I look to continue within the socio-cognitive framework, working very much across the borders that are usually drawn between the social and the psychological. I then turn to document the corresponding continuities between eugenics past and newgenics present. The subject matter here is neither the origins of our ideas of human variation (chapters 5 and 6), nor their manifestation in contemporary newgenics and the continuity between eugenics and newgenics (chapter 7), but the persistence of old-fashioned eugenics itself.

This is very much a question about the persistence of eugenics in the second half of the twentieth century, partly in the wake of the Nazi atrocities in the name of eugenics whose scope and nature received much public and international attention between 1945 and 1950. Yet it might perhaps better be expressed as a question not so much about the shadow cast by those atrocities as one about the continuation of the business of eugenics with seeming indifference to that part of the eugenic past. After 1950, eugenic sterilization continued in a handful of American states, in the provinces of Alberta and British Columbia in Canada, and in all four Scandinavian countries: Norway, Sweden, Denmark, and Finland. All of these states, provinces, and countries were democratically governed, many with substantial levels of social welfare, and none involved explicit eugenic policies of euthanasia or merciful death. The question here is not one about the extremes of eugenics, but about the routinized, normalized forms that eugenics subsequently took.

Therefore, one way to express the question guiding the present chapter is this: Given the scientific, epistemic, and moral limitations and failures of eugenics that we now recognize, how did eugenics continue on as it did? To put it the other way around: How did the social mechanics of eugenics continue to operate, given that there is widespread agreement now that eugenics was fatally flawed on scientific, epistemic, and moral grounds? Those questions are especially striking in the context of eugenics in Canada, given that people were still being sterilized on eugenic grounds right up until the repeal of the Sexual Sterilization Act of Alberta in 1972.

8.2   Subhumanizing Tendencies and Procedural Indifference

My experiences with Alberta eugenics give these questions a twist that deflects several standard responses to it. Since our knowledge of the history of eugenics in Alberta was significantly enhanced through Leilani Muir's landmark legal case that, in turn, spurred many hundreds of related law suits to be filed, a lot of information was accumulated about the social mechanics of eugenics—how people were admitted to institutions such as the Provincial Training School, what written and other information decisions to sterilize them were based on, attitudes and activities of Alberta's Eugenics Board members, negotiations over compensation, and government resistance to that compensation. But most importantly we have the detailed, first-person accounts of eugenics survivors themselves, which tell *their own stories*, stories of what happened to each of them, as they remember it, and that make standpoint eugenics possible. In the present context, there are two significant and intersecting aspects of eugenics as it was implemented in Alberta, both of which have informed the views of eugenics articulated to this point.

The first aspect concerns what, after chapter 4, we might call the *subhumanizing tendencies* that pervade eugenics' constituent practices. These range from the institutionalization and sterilization of children as "mental defectives" without due regard for whether they were properly classified as such, to the psychotropic drug experimentation performed on those children as teenage and preteenage residents in a residential training school.

The second aspect concerns what might be thought of as forms of *procedural indifference* that again pervade the various activities through which eugenics was implemented. Included here are laws that are at best only partially adhered to by the Alberta Eugenics Board (e.g., the sterilization legislation), and a disregard for then-contemporary scientific knowledge, such as findings on environmental influences on IQ test performance. My suspicion is that despite the various ways in which Alberta's eugenics program was distinctive, this subhumanization and procedural indifference is widespread and can be found in many implementations of eugenic programs. But that suspicion is not one that I shall look to substantiate or rest on here.

These subhumanizing tendencies and forms of procedural indifference highlight that the failures we see in the eugenic past are neither simply a function of our putatively superior knowledge—about disability, about genetics, about reproductive technologies—nor of more inclusive and respectful values. This makes unsatisfying two standard and familiar general responses to the question of how eugenics did its work in the post-Holocaust eugenic past.

The first such general response involves an appeal to epistemic ignorance about the workings of eugenics. For example, it might be thought that eugenics continued for as long as it did chiefly because members of the general public didn't know about its detailed workings, or because those responsible for overseeing the implementation of eugenic practices (such as sterilization) were not aware of or didn't realize the marked departures that were made from the appropriate legal and moral norms of the time. To be sure, there is a lot of ignorance manifest in Alberta's eugenic past—for example, about science, about intellectual disability and disability more generally, and about the law. But a core part of what needs to be explained is not how eugenics operated in the shadow of ignorance but in the light that knowledge provides. The fact is that at least *hundreds* of people who at best only questionably possessed the designated eugenic traits were identified as eugenic targets, and institutionalized, trained, and sterilized accordingly. There was an abundance of information available—in individual files, through interpersonal interactions, and in established practices—that showed ongoing eugenic activities and practices to be problematic on their own terms. In addition, part of what needs to be understood is how what ignorance there was—for example, among Alberta's Eugenics Board members—became acceptable and routinized in the continuing micropractices that identified, classified, institutionalized, and sterilized thousands of people over a period of more than forty years.

The second familiar response that is made untenable by the basic facts about Alberta's eugenic history addresses the question of how eugenics continued to operate as it did by invoking a shift in values from the eugenic past to the present. Those in the eugenic past did not value human diversity and disability as we do today, and had their own prioritized social values. Reminding us of the dangers of what is sometimes called a "Whiggish approach" to the historical past, an approach that can only view what happened through lenses shaped by our present values, this response invokes a kind of cultural relativism between eugenic past and the present. But again, although our social values clearly *have* shifted over the past nearly fifty years, part of the puzzle is to understand the operation of eugenics with the subhumanizing valuation of the lives of those most directly affected by it squarely in view. There was much in the values of the day that led to the cessation of eugenic practices in the 1950s—as indeed happened in most North American jurisdictions. But that did not happen universally, and it did not happen in Alberta. Why not?

In short, appealing to ignorance or value shifts does not get to the heart of the question of the persistence of traditional eugenics in Alberta, and I suspect elsewhere. We need something more.

8.3 An Appeal to Wrongfulness

One answer to this question about the persistence of eugenics that is worth exploring, I shall suggest, is rooted in a conceptualization of eugenic ideas and practices as *wrongful*. That particular conceptualization assimilates eugenics to our paradigm of such practices, those that constitute legally wrongful accusation, incrimination, conviction, and nonexoneration. In short, I will be developing the idea that thinking about eugenics by appealing to the model of criminality in the legal system provides some key insights into the psycho-social mechanics of eugenics, into how it is that eugenic practices continued on in the face of dissonant facts and values.

While that idea might be thought original, it in fact derives from a passing comment made by eugenics survivor Ken Nelson more than twenty years ago in the documentary film, *The Sterilization of Leilani Muir*. Ken says, at one point, in describing his life experience as a resident at the Provincial Training School for Mental Defectives in Red Deer, Alberta, that it "almost felt as if you had committed a crime." The children labeled mentally defective, and institutionalized and sterilized on that basis had not, of course, literally committed any crime (or any relevant crime). But many of them—Ken being one of them—also had not metaphorically committed any crime, for they were not in fact mentally defective. They had been *wrongfully* labeled, categorized, subhumanized, sterilized, educated (or "trained"). As their stories make clear, their life prospects were irrevocably stamped with their eugenic past, even as they actively worked to restore their engaged individuality both in their institutionalized and post-institutionalized lives. Part of what is to be explained is the persistence of this kind of wrongfulness.

It is worth clarifying what I am claiming here and what I am not. I am suggesting that we take Ken's standpoint report of what it felt like to be in an institution for the mentally defective when he was not, in fact, mentally defective, as a guide to answering our question about the persistence of eugenics in post-1950 Alberta, in the first instance, and perhaps more generally. That involves taking seriously the idea that eugenics in such circumstances involved wrongful accusation, a notion taken from a legal context. This does not imply that the only thing problematic about eugenics is that it instances wrongful accusation, or that the wrongfulness of eugenics that I am ascribing identifies the greatest wrong that eugenics did. In fact, I think that both of those claims are false, though I shall not argue for this view here. What it does imply is that we should be able to gain some insight into our question about the persistence of eugenics by exploring what we might think of as the social mechanics of wrongful accusation.

Since the concept of wrongfulness that I am relying on is drawn from legal contexts and applies in the first instance to criminal convictions, it is worth saying more about those contexts here. A wrongful conviction in a legal criminal case occurs when the decision to convict someone is mistaken. Initial accusations may be false, police investigative work may be misleading, defense counseling may fail to satisfy minimal standards, or there may be communal pressure to convict the person. The presumption is that the legal system is designed to ensure a fair process in a criminal trial, and that people are assumed innocent until proven guilty. Thus, to move from innocence to a verdict of guilty without the person having committed the crime in question, there must be at least one process leading to the conviction that "goes wrong" in some way. The errors here might be willful and deliberate—as when somebody plants evidence or lies in order to make it appear that someone committed a crime that she did not commit. Or they may be happenstance and accidental—as when blood samples are mistakenly analyzed or crucial documents are overlooked or misinterpreted.

The failures implied by a wrongful conviction are both procedural and systematic. Since the legal system has procedures aimed at preserving the presumption of innocence, in cases of wrongful conviction we say that the legal system failed to deliver justice. Likewise, to attribute wrongfulness to a specific eugenic practice, such as the decision to admit and retain someone in an institution for feeble-minded persons, points to procedural and systematic failures in a larger set of eugenic practices. Although criminal *accusations* are not usually thought to be wrongful in the same sense as are convictions, the concept of wrongful accusation parallels that of wrongful conviction. The very procedural and systematic failures that result in a mistaken outcome or decision can also operate so as to allow a false accusation to be made, to be heard, and to initiate legal proceedings.[1]

8.4   The Case of Ritual Sexual Abuse

What was revealed about the eugenic past in Alberta initially—in Leilani Muir's legal action against the Province of Alberta for wrongful confinement and sterilization—was just this kind of systematic failure. Yet what soon became clear was that these failures were not systematic simply in Leilani's case, but in hundreds (if not thousands) of other cases. People were routinely admitted to the Provincial Training School without sufficient reason for admission; there was frequently either the absence of evidence of their meeting the criteria for sterilization, and often there was positive evidence that those recommended for sterilization did not meet those criteria; and psychotropic drug experimentation on children resident at the Provincial Training School was common practice. These failures were not simply sequentially systematic through the life of a particular individual, but synchronically systematic through the lives of many individuals. The systematic failures documented in Leilani's case were themselves systematic in a broader sense: they represented eugenic business as usual.

For this reason, and with the concept of wrongful legal accusation in mind, our paradigm needs to be not simply an individual miscarriage of justice, however egregious, but wrongful accusations whose systematic nature invades the lives of many individuals. The increasingly well-documented systematic nature of U.S. policing violence, as well as the related overall functioning of its incarceration system—both strongly racialized—represents one kind of paradigm here. But the anchoring paradigm that I want to use is one from which we have a little more historical distance, and I think has closer parallels to the persistence of eugenics. It is the satanic or *ritual sexual abuse* cases prevalent in North America during the 1980s and 1990s.

These ritual sexual abuse cases were a series of cases spanning a period of just under twenty years, and although they occurred worldwide, they were especially concentrated in North America. Beginning with the McMartin preschool case in California in 1983, over the next fifteen years more than one hundred similar cases featuring allegations of ritual sexual child abuse, often involving "satanic" elements, developed in the United States and Canada. These cases included extensive charges of multi-child sexual abuse involving many daycare workers (as in the McMartin case) or groups of largely working-class parents (as in the Bakersfield case). Originally described as *satanic* sexual abuse cases, these cases quickly shifted to be known as *ritual* sexual abuse cases. Such cases involved the supposedly coordinated actions of many individual adults, so-called pedophilic sex rings that allegedly perpetrated repeated sexual offenses on many individual children, and even on groups of children. Thousands of people were charged in these cases; many hundreds were convicted and served sometimes extensive prison sentences. A relatively small number of people convicted of these crimes—crimes not only that *they* did not commit but that did not even happen—remain in prison to this day.

There are striking features of ritual sexual abuse cases, beginning with the fact that *none* of the alleged pedophilic sex rings existed, and *none* of the people accused, charged, convicted, and imprisoned for their alleged perpetration of ritual child sexual abuse in fact committed the crimes of which they were accused and convicted, and in many cases for which they were imprisoned. In addition, many of the cases involved truly bizarre allegations. These include the claim that the accused flew around in the air unaided, and that the putative victims of the abuse had been pierced with sharp objects although they showed no visible marks of such lancing. It was alleged in some of these cases that babies were (ritually) sacrificed, and that there were secret rooms and underground tunnels in readily examinable locations that were the site of the alleged abuse. Rather than such accusations being dismissed or even simply manifestly called into question in virtue of the bizarre nature of these allegations, the lack of credibility of witnesses, the absence of physical evidence, and the large-scale inconsistencies in the charges, they accumulated force and momentum over time. Ritual sexual abuse cases typically began with a small number of charges, later expanding through investigation to a massive multiplication of charges and extensive legal proceedings. For example, in Wenatchee in the state of Washington, there were eventually approximately thirty thousand charges laid against more than forty people in legal proceedings that lasted for more than three years. Those charges began with a single set of allegations from a pair of preadolescent sisters.

In saying that none of the documented ritual sexual abuse cases involved ritual sexual abuse (or even *any* child sexual abuse, so far as I know), I am not denying, of course, that there may be ritual sexual abuse cases. There may be. Still less am I saying that child sexual abuse does not occur. It does, and it is one of the most reprehensible of all crimes. In fact, the background prevalence of *actual* pedophilic abuse, the moral gravity of that pedophilic abuse, and the strong feelings that it properly arouses, I shall be suggesting, are instrumental in the ensuing psycho-social dynamics governing the *wrongful* accusations, charges, convictions, and imprisonment in the ritual sexual abuse cases. These facts about pedophilic sexual abuse not only primed specific wrongful accusations of ritual sexual abuse, but also were structuring causes of the phenomenon itself.

That causal relationship provides the basis for a key insight into the psycho-social dynamics that governed the persistence of eugenics, in Alberta and elsewhere. Furthermore, despite there being legal and social systems that should have detected false accusations of ritual sexual abuse and prevented them from cascading through to wrongful convictions, those systems themselves seemed to foster that very process. Again, the parallel to eugenic practices is perhaps most suggestive. For in cases such as that of Leilani Muir, as well as many of the more than eight hundred cases that were settled by the government of Alberta in its wake, much the same is true. What should have been detected as a false diagnosis of mental defectiveness, or as a mistaken admission to a training school for mental defectives, or an unjustified recommendation for sterilization on eugenic grounds, cascaded through to long-term confinement and eugenic sterilization. Moreover, it did so seemingly in virtue of the very system of protections and checks put in place to prevent so-called innocent children—those who were not mentally defective—from being targeted by eugenic practices.[2]

8.5   Beyond Moral Panic, Groupthink, and Evil's Banality

Within the general socio-cognitive or psycho-social framework that I am assuming, phenomena like those of the wrongful accusations, charges, convictions, and imprisonment generated in the ritual sexual abuse cases are sometimes understood by an appeal to one or another concept in a cluster that appeals to group-level cognitive failures, concepts also useful for understanding kindred cases. These phenomena are sometimes said to involve *moral panics* or *collective hysteria* spreading through a community, or *groupthink* that locks key individual players in group decision making into a sort of tunnel vision from which they can't (or at least don't) escape. Likewise, the concept of the *banality of evil*—Hannah Arendt's characterization of the routinization of death in Nazi concentration camps and within the broader extermination program that these formed a key part of—regularly makes an appearance in initial efforts to understand what happened in the ritual sexual abuse cases.

There is something right about such appeals. The spread of concern about putative ritual sexual abuse is a kind of moral panic, with high levels of emotional reactivity driven by slender threads of evidence. To even so much as raise questions about whether such collective hysteria might exist, at the time, simply led to one of three reactions: either being dismissed from serious consideration, being seen to be complicit with the putative evil being propagated, or even to being accused of participating more actively as a perpetrator. Such reactions stem from groupthink, a tendency for groups to become so rigidified by, and committed to, a specific line of thinking or avenue of inquiry that their collective-level processing becomes not only inefficient and suboptimal, but also contrary to the very aims and objectives they have as a group. And what in other contexts would almost certainly strike one as inhumane, unjust, and a serious violation of people's rights, becomes unremarkable, little more than a part of the everyday social mechanics governing the detection and prevention of a societal evil.

Yet while each of the concepts gets at something important about the phenomena, each is also little more than a placeholder for a deeper psychosocial explanation. Each does something more like simply naming the phenomenon to be explained, rather than explaining it. In this way, resting one's understanding of the social mechanics of the persistence of eugenics on concepts such as moral panic, collective hysteria, groupthink, or the banality of evil contributes to what elsewhere the developmental psychologist Frank Keil and I have called the *illusion of explanatory depth*. This is the sense that one has quite a rich grasp of how something works—whether it be a flush toilet, a car engine, or a psychological agent—when in fact one's ability to provide the sorts of details about the corresponding mechanics or dynamics is, often to one's own surprise, exceedingly limited. Typically, the explanations that we offer for how even familiar, everyday things function or operate are much shallower than one might expect, an observation that Keil and his colleagues have buttressed with a range of interesting developmental studies and that the cognitive scientists Steve Sloman and Philip Fernbach have recently utilized in a popular book on the collective nature of intelligence.

By contrast, the conceptualization of eugenics as involving wrongful accusation opens up new ways of thinking about the social mechanics of eugenics by moving beyond such concepts in a very specific way. It takes us from a recognition of the highly morally and emotionally charged context created by the ideology of the eugenics movement that is embedded in concepts such as *moral panic* or *collective hysteria* to an account of how eugenics embraced subhumanization and indifference in practice. It does so in part because a powerful model for thinking about complicity and witnessing in nonritual cases of sexual abuse also sheds light on cases of wrongful accusation, and so, in turn, on the wrongful accusation model of the psycho-social mechanics of eugenics. This is the dynamic, three-agent model involving victims, perpetrators, and bystanders developed in the classic feminist work on father-daughter incest and sexual abuse more generally by the psychiatrist Judith Herman (1992). To be sure, the use that I will be making of Herman's model, particularly the transformative role of bystanders, would almost certainly be rejected by Herman herself, as we will see. But first, let's review the three-agent model itself.[3]

8.6   Herman on Witnessing and Complicity

Herman's groundbreaking book *Trauma and Recovery* offers an integrated account of two prima facie very different kinds of people living in the aftermath of violent episodes in their lives: soldiers with combat experience, originally recognized as (chiefly) men suffering from "war neuroses," and people (largely women) who had been subject to sexual abuse, especially childhood sexual abuse, and were originally dismissed as hysterics who had largely fantasized the abuse that they reported. Herman brings the two together through conceiving of them as *survivors of trauma*, with the experience of the distinctive trauma in each case beginning a long-lasting trajectory in the downstream life of the survivor. *Trauma and Recovery* itself has been instrumental in consolidating the acceptance of *post-traumatic stress disorder* (PTSD) as a clinical condition, which was first recognized in the third edition of the *Diagnostic and Statistical Manual for Mental Disorders* in 1980. Herman's work has also played a central role in establishing a key place for therapeutic treatment for PTSD in the lives of both men and women.

Herman starts her introduction to *Trauma and Recovery* by reminding readers that the "ordinary response to atrocities is to banish them from consciousness," going on to highlight the destructive nature of the silence and secrecy that follows from this ordinary response for the victims of those atrocities, principal among which are crimes of childhood sexual abuse. The dialectic of trauma that ensues, which leads a victim of such an atrocity to oscillate between the shame of secrecy and the need for acknowledgment of the reality of what happened, is also shared, says Herman, by those who are *witnesses* to those atrocities, where such witnesses include those in whom survivors confide their stories.

The central notion of witnessing in this extended sense is taken from feminist political activism that emphasizes the roles of solidarity, consciousness-raising, and community building as key elements to sexual and gendered emancipation. Herman's introduction of the concept of witnessing at the outset of her book lays the foundation for a powerful model not simply for sex-based trauma, but for thinking about sexual crimes more generally. This dynamic model draws on Herman's experience as a psychiatrically trained therapist for victims of sexual crimes, experience that more directly informs her earlier book, *Father-Daughter Incest*.

The model and Herman's articulation and deployment of it have been extremely influential in shaping the direction of academic, legal, and activist work related to sexual criminal behavior. In this model, in addition to the victim and the perpetrator of this type of crime, there is a third agent, the witness or observer. As Herman writes near the beginning of *Trauma and Recovery*:

To study psychological trauma is to come face to face both with human vulnerability in the natural world and with the capacity for evil in human nature. To study psychological trauma means bearing witness to horrible events. When the events are natural disasters or "acts of God," those who bear witness sympathize readily with the victim. But when the traumatic events are of human design, those who bear witness are caught in the conflict between victim and perpetrator. It is morally impossible to remain neutral in this conflict. The bystander is forced to take sides.

Thus, rather than there simply being perpetrators and victims, there is a more complicated dynamic in play, one among the perpetrator, victim, and *bystander*. As Herman makes clear in the preceding excerpt, the bystander is a potential witness to the traumatic events, as they are recounted by the victim in the path from victimhood to survivorship and recovery.

As the quotation makes clear, part of that dynamic involves the choices and actions of this third agent, the bystander, including those choices and actions that shift him or her from the position of bystander to that of witness. In the next section I focus on the role that such agents play in Herman's model and its appropriations in the humanities and fragile sciences, and what this reveals about the phenomena of witnessing and complicity in sexual crimes. In section 8.8, I turn to what this model tells us about the wrongful accusations of sexual criminal behavior that one finds in ritual sexual abuse cases, and in the case of eugenics.[4]

8.7 From Innocent Bystander to Ally and Advocate

A bystander is someone who is not an active participant in a given action or event, yet still has some kind of presence with respect to it. In the sexually violent cases that Herman has primarily in mind, this presence rarely obtains by a bystander's being there at the scene at the time. Instead, presence is generated through a post-event narrative, one in which both victim and perpetrator have some stake. As Herman makes clear, whatever nonparticipatory innocence that such a bystander has is necessarily lost, since the "bystander is forced to take sides," the side of the victim or of the perpetrator. Although Herman subsequently depicts the decision that a bystander must make as somewhat of a struggle, it is clear what she thinks is morally and politically required of that bystander, someone who in effect has been called to "bear witness to horrible events." To fail to act, or even to act with hesitation or without commitment, is to be *complicit* in the continuation of the trauma, and perhaps even in the act itself, should the perpetrator remain at large in society.

In the first instance, bystanders are those to whom victims of sexualized crimes turn, and on whom they come to rely for support of various kinds. As such, they are typically individuals in special positions of trust and responsibility—as friends, people in a position of special authority (such as a police officer), or with some other professional role to play (such as a therapist). The very act of confiding a trauma-based episode or series of events often serves to intensify whatever antecedent level of intimacy, trust, and sense of responsibility there is between survivor and witness.

Herman makes it clear that individual- and society-level commitments to victims—to believe them, to stand by them—is a crucial part of what is sought by those surviving trauma. Those commitments are a critical component in the long-term recovery process that allows victims to overcome the helplessness induced by the violence they have experienced, moving from victim to *survivor*. Once she is drawn into the dialectic between victim and perpetrator, a witness is no longer a bystander, but a *stand by-her*. That witness must, as Herman says, choose.

My chief interest here in the three-agent model of trauma and recovery is in understanding the psycho-social dynamics that it invokes. In the first instance, this dynamic unfolds at the level *of individual interpersonal interaction*. It challenges those who might otherwise feel they have a more limited or removed role to play in the process of recovery to see themselves in a certain way. They come to self-identify *as witnesses*, a role encompassing both that of an observer—a kind of eye-witness one step removed, of the traumatic events themselves—and that of a confidant, ally, or advocate of the victim as she moves beyond victimhood to survivorship. That dynamic also signals to victims that they are part of a larger phenomenon: they are not alone. They are not alone because there are many others who have lived experience of the same trauma. And they are not alone because there are those whom they can trust as witnesses to what has happened to them, and who can be relied on to stand by them as they work to become survivors of the trauma. Thus, the psycho-social dynamics in the three-agent model serves to forge a dedicated partnership or alliance between victim and bystander-cum-witness against the perpetration of violent trauma, and against the perpetrator of that violence.

This dynamic also operates, however, for those at further remove from those interactions, at what we might call the *level of background politics*. Consider what are often regarded as the kinds of acts of evil represented by the worst of sexual crimes: cases of bodily invasive, long-term, psychologically traumatizing child sexual abuse, paradigmatically incestuous sexual abuse. Such sexual crimes against children exploit the trust that victims have in perpetrators in an especially egregious manner. In so doing, such sex crimes prey on the victim's physical and psychological vulnerabilities in ways that cause long-lasting damage that other forms of sexual crime may not. As Herman points out, it is not simply victims whose credibility is called into question in the narrative struggle with perpetrators, but also those drawn on as witnesses. Thus, there is a need for a broader, political level of support for both victims and bystanders who come to serve as witnesses. They are to be believed and supported collectively as part of a larger political struggle. This gives those who are at least one step further removed from the perpetrator-victim dynamic an important witnessing role. In effect, this shifts child sexual abuse from a private to a political matter, and contributes positively to the ongoing resistance to the sexual oppression of women and children of both sexes, but especially girls.

The commitments and actions of both types of witnessing—at the level of individual interpersonal interaction and at the level of background politics—require overcoming what might be considered natural tendencies to remain silent or nonparticipatory. Bystander-cum-witness action here is effective both for individual recovery and resolution and for larger-scale social change: not just this victim here and now, but also survivors of sexualized trauma, past and future. There is thus much at stake in that initial, simple-sounding call to act in a specific way: to witness.

In fact, the role here calls for much more activity than the labels "bystander" or "witness" suggest. Both at the level of individual interpersonal interaction and the level of background politics, the role of the bystander is that of an ally of, or advocate for, the victim. For this reason I shall sometimes refer to those occupying that role as allies or advocates.

Herman's dynamic model and what we might call the pragmatics of witnessing have influenced, in many positive ways, our broader culture via their effects on institutional policies, laws, advocacy groups, and practices of loyal friendship, in many positive ways. Recognition of this is commonplace among professionals who work with victims of sexual crimes, or among academics who study such victims and such crimes. As a consequence, Herman's model is not simply dynamic and normative in its content, but itself stands in a dynamic and normative relationship to the phenomenon that it seeks to model, exemplifying a distinctive form of what the philosopher of science Ian Hacking has called the *looping effect of human kinds*. While Herman's model captures important aspects of the perpetration of sex crimes, it also introduces a normative role for those who are witnesses to the dynamic between perpetrator and victim via their identification as allies and advocates. In introducing that normative role, the model issues a kind of challenge to those who identify themselves as playing that role. The uptake of that role is one way in which the model dynamically influences the phenomenon it models.

While the influence of Herman's model makes this point uncontestable, less clearly recognized, and in many ways, deeply problematic, is the influence that the victim-perpetrator-bystander model has had in cases of *false* accusation with respect to sexual crimes, especially those of wrongful accusation that occurred in the ritual sexual abuse cases. Indeed, Herman herself appears to have been sufficiently absorbed by the proper application of her model to have become swept up by the ritual sexual abuse cases, dramatically manifesting the phenomenon I view as central to understanding wrongful accusation and its propagation. In her afterword to the revised edition of her *Father-Daughter Incest*, Herman discusses the ways in which allies and advocates have become targets in recent and ongoing cases of child sexual abuse. She says:

In a recent, closely watched decision in the state of Washington, a jury supported the authorities who had investigated an organized sex ring in a small rural town. In this case, twenty-eight adults were charged with child sexual abuse; fourteen pleaded guilty, five were convicted at trial, three were acquitted, and charges against six others were dismissed. The three who were acquitted sued the investigators, claiming that their civil rights had been violated. After the jury found that the authorities had not acted improperly, County Sheriff Dan LaRoche expressed relief that his department would be able to continue investigating reports of child abuse without fear of reprisals: "When your livelihood is threatened for doing your job, and your family's jeopardized," he said, "that's pretty hard to take."

Herman complacently passes on without further comment about this case. But the case she is referring to is the Wenatchee case that I mentioned briefly and passingly, and we should be clear about what transpired here.

In that case, allegations beginning in 1992 but becoming more concentrated from August 1994 eventually led to the arrest of forty-three people by April 1995, with (as Herman reports) twenty-eight people convicted. Eighteen of these people were jailed. All told and as mentioned, slightly fewer than thirty thousand charges were filed in this case, and as Herman is aware at the time of writing, the majority of those charged were either acquitted, had charges against them dropped, or the cases against them were dismissed.

What Herman omits to say, and what she shows no sign of recognizing the significance of, is that in 1998 the Innocence Project Northwest, formed by students at the University of Washington, had taken on the cases of the eighteen who were jailed. By 2000, *every single one of those imprisoned had their convictions overturned*. In the meantime, five had served their full prison sentences, many had their own children removed from their custody, and all had become social pariahs in their local communities. Within a few years, lawsuits against the city and state would begin to be settled for six and seven figures in awards and settlements likely totaling over $10 million. The Wenatchee case, as early as 1998, had become regarded by some as the most blatant example of the miscarriage of justice through wrongful accusation, conviction, and imprisonment in a ritual sexual abuse case. Yet all that Herman seems to see here is an example of how allies and advocates themselves become targets of some kind of reactionary politics.

If what I have said about the laudatory psycho-social dynamics of the three-agent model of trauma and recovery is accurate, then an account of the less laudatory psycho-social dynamics operant in ritual sexual abuse cases is relatively easy to provide. That account is, in turn, readily applicable to the psycho-social dynamics operant in the eugenic past.[5]

8.8   The Psycho-social Dynamics of Wrongful Accusation

In ritual sexual abuse cases, the perceived normative demands of the witnessing role created an understandable hypervigilance in allies and advocates at both the level of individual interactions and that of background politics. That hypervigilance, in turn, targeted merely potential perpetrators of child sexual abuse, saw victims that were not there, and socially constructed a phenomenon—ritual sexual abuse—that had no independent reality. As a result, allies and advocates at the individual interpersonal level *became perpetrators,* not of child sexual abuse itself but of wrongful accusation, and more. And those targeted as perpetrators *became victims* of the same. But allies and advocates at the background political level also became perpetrators of wrongful accusation, reinforcing rather than arresting or halting the initial wrongful accusations.

Simply put, in the ritual sexual abuse cases the zealous actions of would-be bystanders-cum-witnesses of actual, morally charged crimes of child sexual abuse—such as social workers, police officers, parents, and other concerned citizens—created new victims: the innocent adults who had committed no relevant crime at all, as well as children who came to falsely believe they had been ritually abused (and were, effectively, called as witnesses to crimes that did not occur). In effect, those presuming to be witnesses of actual child abuse, and thus allies of and advocates for presumed victims, themselves *became* perpetrators of wrongful accusation, charges, convictions, and imprisonment. This perversion of the witnessing or bystander role was a significant part of the psycho-social dynamics of wrongful accusation, with the normative demands of that role themselves being instrumental in the role's perversion. We encapsulate this shift from how the three-agent model operates in actual cases of child abuse to those of ritual abuse in table 8.1.

The same shift depicted in table 8.1 occurs, I am suggesting, in the shift from eugenic ideology to eugenics in practice: the perceived normative demands of ally and advocate roles created an understandable hypervigilance in witnesses at the level of both individual interactions and background politics. Here community leaders and those in positions of scientific or political authority were allies of or advocates for victims of the eugenic crime of multiplying "the evil by transmission of the disability to progeny"—to take the phrasing from the original Sexual Sterilization Act of Alberta. But in addition to this level of individual interaction, there is also a background political level of alliance and advocacy that members of the general public could join in order to help arrest this eugenic crime. In both cases, to intervene in the name of eugenics is to ally oneself with those who would be affected by eugenic crimes: those who are normal in present and future generations. There is, as in the case of child sexual abuse, a moral demand for bystanders to do more than merely stand by: they

Table 8.1

From actual to ritual sexual abuse

Role: Occupant in actual child abuse	Occupant-role shift in ritual sexual abuse	Comment
Perpetrator: adult	Adults *become victims*.	Note the shift to plural occupants of all three roles.
Victim: child	Children *become bystanders*.	Children are also victims of devastating false beliefs (e.g., their parents or care providers abused them).
Bystander: ally or advocate	Advocates *become perpetrators*.	This shift is the driving force of the dynamics in the model.

must become stand-by-hers—allies and advocates—of intergenerational improvement through the decisions and choices they make in the eugenic present.

The ensuing hypervigilance about eugenic traits in effect targeted merely potential perpetrators of eugenic traits—the poor, the disabled, the foreign, the diseased. It too saw victims where there were none, and socially constructed a phenomenon—the menace of the feeble-minded— that had no independent reality. As a result, allies and advocates at the individual interpersonal level *became perpetrators,* not of eugenic traits like feeble-mindedness but of wrongful accusations directed at those putatively possessing such traits. And those targeted as perpetrators of eugenic traits *became victims* of those accusations. But allies and advocates at the background political level also became perpetrators of this wrongful accusation, reinforcing rather than arresting or halting the initial wrongful accusations. Paralleling the dynamics summarized in table 8.1, we have those I am postulating in the case of eugenics, letting "the feebleminded" stand in for those deemed to have eugenic traits, and "the normal" for those supposedly affected by the crime of eugenic transmission (table 8.2). As was the case with the shift from cases of actual to ritual sexual abuse, here the activity of advocates in becoming perpetrators looms large in driving the social dynamics in play.

This account of the social mechanics of the persistence of eugenics in at least the Alberta context rests on two claims. First, that the three-agent model has a proper or appropriate application to the case of eugenics, and second, that the departure from a desired or idealized mode of operation of that model is central to understanding the social mechanics of eugenics-in-practice.

On the first claim, recall that the proper or appropriate application of the three-agent model in the case of sexual abuse is one in which the

Table 8.2
From eugenic ideology to eugenics in practice

Role: Occupant in eugenic ideology	Occupant-role shift in eugenics in practice
Perpetrator: the feebleminded	The feebleminded *become victims.*
Victim: the normal	The normal *become bystanders.*
Bystander: ally or advocate	Advocates *become perpetrators.*

beliefs of witnesses that sexual abuse has actually occurred are true. In such cases, childhood sexual abuse paradigmatically has an adult (male) perpetrator, a child victim, and a professional ally or advocate. In the case of eugenics, the corresponding circumstances are ones in which the beliefs of allies and advocates that *there really is a eugenic threat* are true. In such cases, eugenic traits paradigmatically are possessed by adults, who are perpetrators of those traits, those deemed normal in present and future generations are victims, and scientific and political authorities are allies and advocates.

For the second claim to be true, the wrongfulness of eugenics must involve a shift in how these roles functioned. In particular, those taking on the role of advocate—and in so doing allying themselves with potential victims of a kind of eugenic crime—became perpetrators of wrongful accusation. And those targeted as mental defectives—people like Leilani Muir, Judy Lytton, Glenn Sinclair, Ken Nelson, and Roy Skoreyko in Alberta— shifted from being putative perpetrators of defectiveness to becoming victims of that wrongful accusation.

If this is correct, then the persistence of eugenics can be explained by modeling eugenics as a form of wrongful accusation, much like that which existed in examples of ritual sexual abuse cases. If eugenic survivors like Ken Nelson report their feeling that it was almost as if they had committed a crime, it is because that is precisely how they had come to be viewed in society: as those poised to commit some kind of eugenic crime.

8.9 Persistence Redux

My adaptation of the three-agent model to explain the fundamental persistence of eugenics in the postwar era does not (and does not try to) resolve questions about larger-scale social policy and societal values, any more than does understanding the dynamics between real child sexual abuse and the merely putative ritual sexual abuse of children resolve the corresponding large-scale policy decisions there. My purpose here has not been to answer the question of whether the perceived eugenic threat of degeneracy and the degradation of the gene pool justifies policies regarding institutionalization and sterilization that have, as a byproduct, a certain kind of wrongful outcome, or whether that systematic wrongfulness is an unfortunate but on balance necessary evil for the protection of society. The historical repudiation of traditional eugenics has, one might think, provided one set of answers to such questions. Insofar as the argument here contributes to an understanding of a neglected aspect to the dynamics of The Eugenic Mind that leads to the persistence of eugenics, it should, of course, inform our engagement with those larger questions, particularly as they arise in new forms but also as we reexplore the relationships between eugenics past, present, and future.

This chapter has rounded out my discussion of the persistence of eugenics by taking a novel approach to the question of how eugenics continued as it did in the postwar era, given the scientific, epistemic, and moral limitations and failures of eugenics that we now recognize. Having put aside some familiar kinds of answers—ones that appeal to moral panic, to groupthink, to collective ignorance, to the banality of evil, or to shifting values—as lacking the kind of explanatory depth that, I believe, the question demands, the chapter has concentrated on developing a simple but perhaps initially strange idea: that eugenics itself could be understood as a form of wrongful accusation. More precisely, the claim was that we might understand an important part of the social mechanics governing the persistence of eugenics in the second half of the twentieth century by attending to the psycho-social dynamics of witnessing in cases of ritual sexual abuse. Here, as throughout part II, I have looked to integrate hypotheses and approaches from across the fragile sciences—particularly the social and psychological sciences. As with *The Eugenic Mind Project* more generally, the perspective developed here is very much a part of a standpoint eugenics, drawing on the epistemic and moral insights afforded by eugenic survivors themselves.

The core of the discussion in this chapter, however, has focused on articulating a general model for understanding wrongful accusation, conviction, and imprisonment by concentrating on one paradigmatic example, that of the ritual sexual abuse cases prevalent in North America during the last two decades of the twentieth century. Here I have engaged in three activities—describing, explaining, and adapting—none of which involve eugenics per se. First, I have described the salient features of the ritual sexual abuse cases; second, I have explained what I call the three-agent model developed originally by the psychiatrist Judith Herman for understanding father-daughter incest and broadened as part of her advocacy for the importance of *witnessing* in cases of sexual abuse and trauma more generally; and third, I have adapted that three-agent model to the specific case of ritual sexual abuse that constitutes a paradigmatic form of wrongful accusation, conviction, and imprisonment. With that work done, the application to post-1950s eugenics—particularly in Alberta but (I have hypothesized) more generally—is relatively straightforward, and has occupied only section 8.8.

As the name suggests, the three-agent model is a dynamic representation of the relevant relationships among three agents, as the name suggests: a perpetrator, a victim, and a bystander-cum-witness. In cases of sexual abuse—paradigmatically father-daughter incest (for Herman) but more generally—the asymmetry in power between the perpetrator and victim places a moral demand on bystanders to become stand-byers, not simply epistemic and moral witnesses for the victim but her allies and advocates. Herman's model is both descriptive of the dynamics among these three agents and instrumental in directing how those agents, particularly the bystander-cum-witness, ought to behave. In the typical case, not being present as an *eye-witness* (or even *ear-witness*) to the putative crime of sexual abuse, the bystander here is a witness, in the first instance, to the testimony of the victim, and secondarily to the broader context in which the relationship between perpetrator and victim needs to be understood.

The central idea in play here to the adaptation of this three-agent model to cases of wrongful accusations of child sexual abuse—paradigmatically the ritual sexual abuse cases—is that once we understand how the three-agent model operates in what we might think of as the proper functioning cases, those for which it was developed by Herman, we can see how the dynamics of the model itself creates, stabilizes, and reinforces wrongful accusation, conviction, and imprisonment in the case of ritual sexual abuse. The very features that make the three-agent model such a powerful tool, both for understanding *actual* child sexual abuse and for holding those who *actually* perpetrate it accountable, themselves have functioned to fabricate wrongfully alleged child sexual abuse in cases of ritual sexual abuse. Though the sexual abuse in these cases did not actually exist, the effects on the lives of all three agents were all too real. Lives truly were destroyed, not only those of falsely accused perpetrators (who thus became victims), but also those of falsely identified victims (who thus became witnesses themselves to crimes that did not occur). Those newly created witnesses were children who typically came to falsely believe that people whom they loved and in whom they had significant levels of trust and emotional investment—their parents, close relatives, and nonfamilial care-providers—had violated their bodies in the most egregious and disturbing ways. It is for this reason that I have identified them as not only witnesses, but also as victims of false accusation.

This core claim—that the very features that make for effective understanding and action in normal cases also create the phenomenon of wrongful accusation—carries over to the case of eugenics as wrongful accusation. In what I have been calling The Eugenic Mind, there is a dynamic relationship among the same three agents—perpetrators of eugenic traits, victims of the propagation of those traits into future generations of defective people, and those bystanders called to rise as allies of and advocates for those victims. My claim is that, just as in the case of ritual sexual abuse, the wrongful confinements and sterilizations that systematically recurred during the second half of the twentieth century in places like Alberta were promulgated through the very psycho-social dynamics of this three-agent model. What was seen as an aberrant outcome in practice was, in fact, the result of the way The Eugenic Mind itself operated in its proper functioning case.

Although this articulation of the idea of eugenics as wrongful accusation applies, in the first instance, to the wrongful confinement (through institutionalization) and sterilization that was prevalent in postwar eugenics in Alberta, it would be mistaken to view the adaptation of the three-agent model as being limited in application to what some might think of as special, localized circumstances. For that model will apply wherever there are, in fact, *no perpetrators of eugenic traits*—the subnormal—yet eugenic ideology is cast in terms of improved future generations lacking subnormal traits and calls for "bystanders" to ally with and on behalf of those improved future generations—the normal. This will be the case where the traits that specify people as eugenic targets do not in fact pose the transmissible, intergenerational threat that they are alleged to pose.

As a start, we can look back to the list of eugenic traits that were specified in sterilization legislation summarized in table 3.1 to see which traits this might be true of: feeble-mindedness, insanity, epilepsy, criminality, imbecility, idiocy, sexual perversion, mental deficiency, rape, mental disease or illness, syphilis, being an institutionalized person, or pedophilia. Recall that I suggested there that these traits fall near exhaustively into a primary cluster—concerning psychological traits and mental health—and a secondary cluster—concerning sexual behavior, together with epilepsy (which we might integrate into the former cluster). For any one of these traits that does not in fact create intergenerational "victims," sexual sterilization on the basis of that trait is wrongful in the sense I have specified here, and the three-agent model sheds light on the psycho-social dynamics that led to the persistence of eugenics as a form of wrongful accusation.

If this is correct, then the model I have introduced has the potential to make sense of much of the persistence of not only eugenic sterilization but also eugenics more generally. For one might plausibly argue that some of these eugenic traits (such as those concerning sexual behavior) are not genetically transmitted; that some (such as many of those clustered as mental) cannot in fact be effectively prevented through sterilization; and that some (such as epilepsy) do not constitute the menace to society that they have been alleged to. In fact, just those arguments have been made both in repealing sexual sterilization legislation and in more quietly ceasing the practice of eugenic sterilization.

The Eugenic Mind persists beyond practices such as sterilization, and one might wonder whether the three-agent model and the idea of eugenics as wrongful accusation are helpful in understanding this dimension to its persistence. I am thinking in particular of the eugenic strain of thinking that is sometimes explicit, as we have seen in chapter 7, and often implicit in contemporary discussions of prenatal screening and selective abortion and of human enhancement. Consider again Down syndrome to see how the idea of wrongful accusation might apply.

We have seen that proponents and critics of selective abortion here disagree in part about what the prevalent use of prenatal screening to detect fetuses with trisomy 21 and then terminate those fetuses says, means, or expresses about both those fetuses and people with Down syndrome. The idea of wrongful accusation sheds some light, I want to suggest, on the perspective of the disability rights critique of selective termination in these cases. For proponents of this critique, the fetus has come to be viewed as the means through which undesirable traits appear in one's infants and children, with that view being taken to be wrongful either because it expresses synecdoche—with one trait standing in for the value of the whole child—or because it mischaracterizes the lives of people with Down syndrome and the families they belong to. While we might well balk at the extension of the labels "perpetrator" and "victim" here, and so stop short of applying the three-agent model in a literal way to this case, what is true for the critic of selective termination in such cases is that a putative agent of some kind of undesirable trait—the fetus diagnosed as having trisomy 21—has become the target of elimination, and wrongfully so. A model for vigilantly and appropriately protecting visions of family life from severe disease has become misapplied, leading to the elimination of a trait (and its bearer) whose perceived undesirability likely far outstrips the threat that it actually poses to that vision.

III   Eugenic Voices: Knowing Agency at the Margins

9 Knowing Agency

9.1 Marginal Knowing

The perspective that I have developed in parts I and II has given importance to the lived experiences of eugenics survivors for understanding and explaining eugenics, past and present. Here I have drawn openly on my own experiences in working together with survivors of Alberta's eugenic past. Primary among such survivors are those whose lives had been governed fairly directly by eugenic sterilization laws, such as the Sexual Sterilization Act of Alberta. But as I noted in chapter 1, the idea of survivorship in this context is broader, encompassing also people with disabilities in our local community who had come to see their day-to-day lives in terms of very much the same kinds of subhumanization and social exclusion that had been implemented through those eugenic laws and policies of years past.

Both types of eugenics survivors are marginal knowers in multiple senses. First, they have been historically marginalized in society as the sorts of people who were eugenic targets. Second, their perspectives have played, and continue to play, at best a peripheral role in our collective knowledge generation focused on eugenics. This is so despite the engaged individuality of those survivors that, together with that marginalization, provides those survivors with distinctive and distinctively informative views of The Eugenic Mind.

Early in chapter 1 I introduced the general idea of a standpoint eugenics as an expression of the perspective on eugenics that I have developed, one that emphasizes the epistemic value of the standpoint of such marginalized individuals. There I also provided a brief characterization of standpoint theory, the most developed, systematic account of the epistemic importance of the perspectives of oppressed and marginal subjects in the humanities and social sciences and to which the label "standpoint eugenics" pays homage. Throughout *The Eugenic Mind Project* I have proceeded very much by *doing* what I take to be standpoint eugenics, rather than engaging in precisifying discussion of just what a standpoint eugenics must, can, can't, and mustn't be, and the hand-wringing that comes almost inevitably with such discussion. Here I turn to reflect more directly and critically on the very idea of a standpoint eugenics, moving beyond the playfulness invoked by my earlier deferring appeals to a standpoint-ish eugenics.

Thus, in this final part of the book, part III, "Eugenic Voices," I come to focus more explicitly on epistemological issues at the heart of standpoint eugenics, what I call *knowing agency at the margins,* a phrase that refers both to a sort of agency ("knowing") and to the activity of finding out ("knowing") about a certain sort of agency. It is the sort of agency that derives from and produces knowledge about those who are systematically marginalized. Knowing agency at the margins is precisely what a standpoint eugenics aims at understanding in the case of actual and putative disability. Here disability with an intellectual, cognitive, or psychological dimension at its underlying core is central, given the importance of eugenic traits such as feeble-mindedness and insanity in the eugenic past, the focus on conditions such as Down syndrome in ongoing discussions of selective abortion, and the positive selection of intelligence in contemporary defenses of human enhancement.

Standpoint eugenics faces challenges of a mixed sociological-philosophical variety due to its own situatedness as a kind of standpoint theory-in-action, one drawing on the perspectives of a novel type of knowing agent, eugenics survivors. Within philosophy, epistemology (the theory of knowledge) can be divided in many ways, with standpoint epistemology being one approach within epistemology. Standpoint eugenics is located against the background, however, of a kind of epistemic state of separateness, or epistemic apartheid, that exists between traditional analytic epistemology, on the one hand, and standpoint epistemology, on the other. And here as elsewhere, rather than describing the character of that state of separateness, the phrase "separate but equal" merely expresses hopeful or wishful thinking. This is because standpoint epistemology very much has marginal status vis-à-vis traditional analytic epistemology, as I shall attempt to show.

In addition, within standpoint theory, standpoint eugenics must be articulated against a sort of presumption of universalism within standpoint theory itself, with accounts of the standpoints of eugenics survivors being taken, at least by default, to be assimilable to that of other standpoint agents. Yet this presumption itself deserves interrogation, or at least so I shall argue. I begin with something briefly elaborating on the background state of separateness within epistemology.

Traditional analytic epistemology is the dominant approach to the theory of knowledge within analytic philosophy. It is what undergraduates are introduced to as epistemology in most departments of philosophy in North America and Great Britain, and focuses on topics such as whether knowledge can be defined or adequately conceived of as justified true belief; whether truth is best thought of in terms of correspondence or coherence; what the nature of justification is; what skepticism, fallibilism, idealism, realism, and empiricism claim, and strengths and weaknesses of each of these views; and the views of major historical figures, such as Descartes, Locke, Berkeley, Hume, and Russell about the reach and limits of human knowledge. Despite a recent contextual turn within traditional analytic epistemology, political concerns about who knows (and who doesn't) remain relatively neglected.

The most developed epistemic view centered on the situatedness of knowers, and their differential positioning with respect to knowledge, evidence, justification, and truth, is feminist standpoint epistemology, which developed from, or as a part of, feminist standpoint theory more generally in the social sciences, particularly in sociology and political science. Although one might encounter standpoint epistemology in one's introduction to analytic epistemology, as I shall argue further in section 9.2, these two approaches to epistemology remain largely segregated from one another.

Feminist standpoint epistemology within philosophy, however, has shared with standpoint theory more generally the supposition that it could readily generalize from particular types of standpoint agents—particularly the working class and women—to provide a general account of knowing agency from the margins that encompasses a broader variety of standpoint agents. Or at least that has been the implicit working assumption of both standpoint epistemology and standpoint theory more generally. But as I shall argue in chapter 10, this universalist hope is likely to be disappointed, particularly in the case of disability and eugenics survivorship, but more generally, without some significant revision to core parts of classic standpoint theory. By attending to features of standpoint agents more apparent in racism, heterosexism, and ableism than in more commonly discussed cases of sexism and classism, I will propose ways of enriching accounts of knowing agency from the margins that are particularly apt for thinking about nonstandard standpoint agents. Standpointing eugenics not only casts eugenics in a new light, but also reinvigorates standpoint theory itself.

Broadly speaking, the preoccupation in this pair of chapters is with the politics of knowledge at the margins, a politics manifest both in addressing the question of *who* knows, and in reflecting on the state of separateness within epistemology itself. Who cares about who knows is itself as much a part of the politics of knowledge as is who knows.[1]

9.2   Who Cares about Who Knows?

Political concerns about who knows, and who doesn't, and the ways in which values infuse authoritative and trustworthy knowledge, have themselves been peripheral topics in analytic epistemology's own history. The most developed subfield in which such concerns have generated sustained attention is the epistemology of scientific knowledge, attention buttressed by the interdisciplinary field of science and technology studies and subsequently informed by work on gender and science. Despite some prominent, recent work by Miranda Fricker and Laurie Paul more squarely addressing questions at the heart of traditional analytic epistemology, the politics of knowledge continues to linger at the edge of traditional analytic epistemology.

This is neither to say that questions about the contexts in which knowledge claims are made, justified, questioned, or refuted have not been raised within traditional analytic epistemology, nor that their implications for the credibility, authority, and the objectivity of knowledge claims go completely unrecognized. But there is a way in which those questions and implications have resisted or simply been disconnected from a politics tied to action and engagement. While traditional analytic epistemology recognizes context-sensitive approaches to epistemology, it peripheralizes politics-infused versions of such approaches.

One can see this vividly by considering the development within analytic epistemology of a subfield of work that does attend explicitly to various ways in which the context in which knowledge claims are made either affect the truth value or nature of those claims. Despite this focus on the context of knowledge, the body of work produced under the name *epistemic contextualism* remains distant from concerns about the politics of knowledge: how it is produced, by whom, and for what purposes.

As Patrick Rysiew's review article on epistemic contextualism indicates, contextualism about knowledge within the analytic tradition is focused tightly around classic paradoxes and puzzles (e.g., Keith DeRose, Crispin Wright), modifications of naturalistic epistemology (e.g., Fred Dretske, Alvin Goldman), and integration with related views in metaphysics and the philosophy of language (e.g., David Lewis, Stewart Cohen). Looming large in the literature are the "isms" that have dominated traditional epistemology—skepticism, relativism, coherentism, and fallibilism, for example. Virtually absent is a concern with how context interacts with the *subjects* of knowledge, particularly differences between what I prefer to call knowing agents. Indeed, Rysiew's review partially delineates epistemic contextualism in terms of its focus on attributions of propositional knowledge claims and the ways in which context enters into the ascription and evaluation of such attributions, rather than on "certain features of the putative subject of knowledge" (first paragraph). This focus is sufficiently strong to refer to epistemic contextualism as *attributor contextualism*, and I suspect that both Rysiew's delineation of epistemic contextualism and its characterization as attributor contextualism are intended to reflect the reality on the ground of what contextualism in epistemology does, and does not, encompass.

As intimated in section 9.1, such concerns with the politics of differential knowledge have been central to feminist epistemology. These concerns span both the analytic and continental traditions, and have been occasionally taken up within strands of naturalistic epistemology originating within the philosophy of science. The most developed view here, particularly within feminist epistemology and feminist work in the social sciences with which it is intertwined, is standpoint epistemology or standpoint theory more generally.

The central tenet of standpoint theory is that subjects marginalized through at least one system of oppression are sometimes so positioned due to their subsequent experiences that their views and claims have some kind of epistemic privilege. Moreover, the knowledge generated from this standpoint plays a particular, transformative role in resisting and even overthrowing the corresponding marginalizing structures of oppression. Standpoint theory thus identifies marginalized individuals as special sorts of knowing agents. As I noted in chapter 1, standpoint theory originated in Marxist views of the working class and was redeployed to theorize gendered oppression by social scientists before making its way into philosophy.

The relevant epistemic privilege afforded by standpoint epistemology to such knowing agents at the margins isn't simply given once and for all. Instead, a standpoint is best understood as a form of epistemic engagement, "a matter of cultivating a critical awareness, empirical and conceptual, of the social conditions under which knowledge is produced and authorized," as Alison Wylie has said. As a critical form of engagement, a standpoint cultivates and accrues political awareness and subjectivity over time. To "take up" or "occupy" a standpoint is therefore a project in the existential and political sense. It is to realize oneself in increasing measure as a political agent who, as a marginalized other, has been subjectivized to be silent, but nevertheless possesses subversive epistemic privilege.

While standpoint theory is intended to provide a general account of knowing agency from the margins, it has concentrated on certain kinds of standpoint agents, particularly women and members of the working class. Yet there are other sorts of standpoint agents—people with disabilities, for example—who vary significantly both with respect to their individual properties and their social locatedness in ways that problematize a simple generalization of standpoint theory from the cases of gender and class. Before taking up some of the issues that this raises in chapter 10, I want to draw on a sociological observation relevant to the state of separateness between traditional analytic epistemology and feminist standpoint epistemology that I have alleged exists.

As some may have wondered in scanning those named in my brief characterization of epistemic contextualism, yes, the literature in epistemic contextualism is disproportionately dominated by male authors, even given the well-known gender skew reflected in base rates of membership in the society of professional philosophy. For example, in Rysiew's review, more than eighty authors are listed in the references; only five of these are women. Of the roughly equivalent number of authors listed in the references of what might be regarded as the closest comparison case, Heidi Grasswick's article "Feminist Social Epistemology" in the same prestigious and widely read publication, the gender asymmetry is almost as radical, though in the opposite direction; only ten of the authors cited by Grasswick are male. In addition, of the more than 160 authors here, *none appear in both lists of references*, and very few (if any) have published in both areas, despite the fact that both topics have, at their heart, various ways in which context and knowledge interact. Neither article refers to the other, and even the cross-linkage is thin between articles related to each (e.g., "feminist epistemology and philosophy of science" for "feminist social epistemology," and "knowledge: analysis of" for "epistemic contextualism").

There is nothing idiosyncratic about either of these impressive and informative reviews here; their authors have done an excellent job in providing overviews of the topics covered in the literature under each of their headings. The bifurcation between the two articles represents the fact that these distinct ways of approaching the intersection of knowledge and context within the discipline of philosophy are radically disjoint, existing in a state of separateness or epistemic apartheid.

Given their very different standings within the discipline, however, further reflection on these merely sociological differences between epistemic contextualism and standpoint epistemology reveals something further about the politics of *philosophical* knowledge, a topic I shall return to in concluding this chapter. As with my rough yardstick for measuring the state of epistemic apartheid between traditional analytic epistemology and feminist standpoint epistemology, here I draw on just as crude an instrument to bring attention to the differential disciplinary standing of these two ways in which the relationship between context and knowledge have been theorized in philosophy. This simply makes vivid what I think anyone familiar with the relevant journals, fields, and topics already knows.

Citation practices are one important measure of influence in a field. Consider the most highly cited journals in the categories "Philosophy," "Epistemology and Scientific History," and "Feminism and Women's Studies," as indicated by the lists generated by Google of the twenty journals with the highest h-5 indices in each of the categories. In the Google category "Philosophy," the journals listed include (just to take the top ten) *Synthese*; *Philosophical Studies*; *Nous*; *Philosophy and Phenomenological Research*; *Mind and Language; Review of Philosophy and Psychology; Ethics, Phenomenology and the Cognitive Sciences; Mind;* and *Philosophical Psychology;* in the category "Epistemology and Scientific History," journals include *Synthese; Nous; Philosophy of Science; The British Journal for the Philosophy of Science; The Journal of Philosophy; Studies in the History and Philosophy of Science Part B and Part C; Erkenntnis; Journal of Philosophical Logic;* and *Foundations of Science;* in the "Feminism and Women's Studies" category, journals include *Psychology of Women Quarterly; Signs; Women's Studies International Forum; Feminism and Psychology; Hypatia; International Feminist Journal of Politics; European Journal of Women's Studies; Feminist Theory; Feminist Review;* and *Journal of Middle East Women's Studies.* Reinforcing my claim of the state of separateness between traditional analytic epistemology and feminist standpoint theory is the fact that there is *no overlap* between the top twenty h-5 index journals in the third of these categories and those in either of the other two categories.

Supporting the further claim that this difference marks differential status in the discipline of philosophy is that while there is substantial overlap between the first two lists (*Synthese, Nous, The Journal of Philosophy, Philosophers' Imprint, The Journal of Philosophical Logic,* and *Erkenntnis* are on both full lists, for example), there is *no overlap* between those listed in the "Philosophy" and "Feminism and Women's Studies" categories. In fact, the only journal clearly identifiable as a philosophical journal in the latter list is the journal *Hypatia,* which ranks just outside the top twenty h-5 journals in the discipline of philosophy.

To get some idea of how central contextualism is to both traditional analytic epistemology and the discipline of philosophy more generally, and contrastingly how marginal standpoint epistemology is, consider again a crude measure that indicates what one finds in the journal at the top of these rankings: *Synthese.* "Contextualism" returns 141 hits for articles in the journal dating from 1978 to 2016. By contrast, between them "standpoint epistemology" and "standpoint theory" return six hits; "feminism" itself, which encompasses much more than feminist standpoint theory, returns only seventeen hits. My hypothesis is that this pattern would be repeated, were one to examine the content of each of the respective twenty journals under the "Philosophy" heading in turn.

While these sociological differences are readily apparent in the case of standpoint and gender, I also want to draw attention to the less commonly discussed cases of knowing agency from the margins, such as those involving disabled people and indigenous people, whose presence in professional philosophy, let alone within traditional analytic epistemology, has been barely registered. If the knowing agents of feminist standpoint epistemology are very much at the margins of mainstream epistemology, the knowing agents of disability studies have yet to find a place on that same map. Whatever state of separateness from traditional analytic epistemology that standpoint epistemology in general has, forms of standpoint theory that focus on *disability*, real and perceived, must grapple with the further and deeper marginalization of disability itself within the discipline of philosophy. To provide some sense of *what it is like* to occupy this space, as well as to continue to illustrate what it is to do standpoint eugenics, I turn in section 9.3 to some important work—philosophical, community building, and reflective—by Eva Feder Kittay.[2]

9.3   Cognitive Disability and Its Challenges

In September 2008, Eva Feder Kittay and her colleague in philosophy Licia Carlson organized a conference in New York City called "Cognitive Disability: A Challenge to Moral Philosophy," the talks and discussions of which were captured on video, and which later became the basis for a special issue of a journal, as well as an edited book that derived from that special issue, published the following year. As advertised, the conference set out to "explore philosophical questions about three specific populations—people with autism, Alzheimer's disease, and those labeled 'mentally retarded.'" It featured an all-star lineup of moral philosophers, including Peter Singer and Jeff McMahan, whose views of cognitive disability, as mentioned briefly in chapter 4, had been seen as subhumanizing and more generally had raised concern among a number of philosophers of disability, including Anita Silvers, Adrienne Asch, and Leslie Francis, as well as the conference organizers, all of whom were in attendance.

Over on the other side of the continent, we had relatively recently started an active interdisciplinary blog called *What Sorts of People Should There Be?* in what turned out to be an early precursor to the Living Archives on Eugenics in Western Canada project. We contacted the conference organizers to see whether our blogging about the content of the talks in a systematic way, using video excerpts from the conference, might be both possible and welcome, and it turned out to be both. So away we went, finding relatively short clips of the participants, and providing some framing for discussion of them, and some critical thoughts. We ended up writing thirteen blog posts, featuring eleven authors, each post between 1,000 and 5,000 words in length. The discussion the posts generated, both on the blog and beyond it, overshot our expectations in both quantity and quality. At one point, we were attracting more than five times the number of regular visits to the blog per week. I remember thinking at the time that even some of the detailed comments that were made on the posts surpassed, in depth and quality, even quite good papers that I had recently graded from upper-level undergraduate and graduate classes in the area.

The raw materials that made this possible, of course, were the original conference talks and interactions, and they were interesting both philosophically and sociologically. Here we had a very heartfelt and personal meeting point of two perspectives, one from mainstream ethics, the other from the philosophy of disability. Although the kind of difference between these perspectives is a theme that has surfaced several times in the book so far, and there is much more that could be said about it, here I want to home in on the corresponding meeting point in *epistemology*, discussing not so much values and moral theory, but knowledge and its systematic study.

Most relevant here are some comments that Kittay herself made in her paper that offered some explicit reflections on the conference framed around her "positionality": as a professional philosopher who regularly encounters views of cognitive disability that dehumanize people such as her dependent daughter, Sesha. Yet it is not simply those views that represent Kittay with more than a challenge here, but the field's complicity in the acceptance of those views. After leaving one conference in which she had endured an especially frustrating exchange, Kittay reflects: "What, I wondered, was I doing in a discipline that thought it appropriate to question the full worth of a portion of humanity, one that happened to include my own daughter? A discipline whose practitioners sat on the sidelines as I fought to defend her moral worth and that of those like her?" Kittay's answer is as brief as it is striking, drawing on notions of care and the labor of love that occupy a central place in her broader philosophical work:

My daughter, Sesha, will never walk the halls of academe, but when what happens within these halls has the potential to affect her, then I as an academic have an obligation to socialize academe to accept my daughter. Such "care" may seem to be far from the daily care that her fully dependent body requires, and it may appear to be far-fetched to call this "care," but it is part and parcel of that labor of love that we do as parents, especially parents of disabled children—more still in the case of those who are so disabled that they cannot speak for themselves, a defining condition for those who are severely intellectually disabled.

To round out this introduction to the margins of knowing agency at the margins, I provide three vignettes that tell us something more about how the voices of eugenics survivors and others occupying a standpoint eugenics have been received in the professional worlds of academia and the health sciences more generally.[3]

9.4   Vignettes and Voices

Survivor Knowledge
A few years ago, a colleague working with the Living Archives on Eugenics in Western Canada project, not herself an academic but working with one of our community-based partners, returned with the following story from a recent conference of a professional society of historians that meets annually. She had given a talk there about the work that she was doing with Living Archives. As part of her response to the simple question of how the conference was, she said, with her usual sense of irony, "Well, it was great, except this guy there thought that what we were doing was worthless." What does she do? She uses video interviews with eugenics survivors, some of whom have intellectual disabilities and some of whom do not, to build oral histories of the lives of those survivors, histories that provide an insider's view of the history of eugenics, which formally ended in our locale only in 1972. "Why did he think the work was worthless?" I asked with simplistic, knee-jerk curiosity. The reply: "He said that if morons and lunatics telling their own stories provided a reliable way to learn about the eugenic past, what would historians like him have to offer." A clever, self-effacing joke of a kind familiar to academics, but beyond which lies a skeptical answer to the question of the place of the voices of eugenics survivors in the study of The Eugenic Mind.

Parental Advocacy

Through my interest in reflecting on the connections between the eugenic past and contemporary views associated with the sorts of people who were either the explicit or implicit targets of eugenic laws and policies, I have met a number of parents who have become disability or child welfare advocates. They have done so primarily in light of their experience with doctors, hospitals, state and provincial agencies, police departments, politicians, ombudsmen, and others involved somehow or other in the often-torturous struggles they have been through to answer one simple question that each has asked: "Why was my child not thought worthy of life?" In some cases, that question arose as a result of the death of their child in suspicious or questionable circumstances; in others, as a result of negligence, differential treatment, or indifference for which the unworthiness of life appeared to them, eventually, the most plausible explanation. While many of these parents are well-educated and middle-class, none of them are academics (let alone philosophers), and few had been involved in any serious kind of antecedent advocacy work—that is, antecedent to their experiences that almost inevitably became struggles as people parenting around disability. The treatment of their children, and the long and painful back-and-forth between inquiry and institutional or personal response formed them as parental advocates. Their children had various disabilities or genetic conditions indicative of various forms of disablement. This nobody questions. But their shared question—"Why was my child not thought worthy of life?"—nobody answers.

In one case, a young mother found herself in Edmonton where her "severely disabled" infant was receiving hospital-based care. She was staying at a facility for family members in just the kind of situation she was in, about a seven-minute walk from my house in one direction, while her infant was in a hospital about the same distance in another direction. Doctors and other medical staff had come to the conclusion that the infant had irreversible brain damage and had engaged the mother in discussions about end of life scenarios, including a DNR or "do not resuscitate" order and the removal of life-sustaining medical support. Since the case had received some brief coverage in the local press, I knew something general about the situation, but most of the details I was to learn from a phone conversation, not with the mother, but with one of her parents, who was looking, obviously in some desperation, for someone at arm's length from the medical professionals who were providing care, advice, and comfort. As a family, they needed to take stock. There was one point in the conversation where I said that I knew of a case in Ontario where a parent had been removed as the legal guardian for her child *because she had not agreed to have life support removed in the provision of care for her child*. There was an audible gasp at the other end of the line, not because of the shock of the content of what I had said but, it turns out, because of its familiarity. "That's just what my daughter's doctor threatened to do" was the next thing said from the other end of the phone. Life-worthiness and parent-worthiness, it seems, are sometimes related in ways that are hard to believe.

Knowing Agency

Leilani Muir's recently published memoir, *A Whisper Past*, tells Leilani's own story in detail, the first book-length memoir by a eugenics survivor from the Provincial Training School for Mental Defectives in Red Deer, Alberta. The part of the story that I want to highlight here concerns Muir's important role in the history of Canadian eugenics, some of which was already noted in chapter 1. Muir had been institutionalized at the Provincial Training School at the age of ten, and sterilized four years later, putatively in accord with the Sexual Sterilization Act of Alberta. Starting in the late 1980s, and having only discovered nearly ten years after the fact that she had been sterilized, Muir began to seek legal recourse for having been wrongfully confined and sterilized under the Act. In 1996, she became the first (and only) person to successfully sue the province on those grounds, a landmark case that paved the way for more than eight hundred other cases that were settled eventually out of court after much resistance by the government of Ralph Klein. This resistance included the introduction of little-known Bill 26 to the floor of the legislature, which sought to limit payments to eugenic survivors regardless of what had been done to them, a bill introduced *while the government was in settlement discussions with lawyers for the plaintiffs in the Alberta court system.*

Once introduced to a larger slice of the history of eugenics in Alberta, many people ask and sometimes speculate about the question "Why Alberta?" A question asked less often, but still occasionally, is one that has loomed larger in the present book: "Why Leilani?" How could this woman, deemed wrongfully to be mentally defective and sterilized after review and approval from Alberta's eugenics board, with many other antecedent cards stacked against her, and much pre- and post-institutional trauma in her life, come to have such a significant impact on the history of eugenics in Alberta and our knowledge of that history? How did Leilani come to be a knowing agent of such force and impact, one without whom the history of eugenics in Alberta would have remained as buried in 2015 as it was twenty years earlier in 1995?[4]

9.5   Epistemology Impoverished and Knowing Agency

These vignettes and the voices that make them possible raise questions about knowing agency from the margins, questions that traditional analytic epistemology has seldom addressed. These include questions about *who* has authoritative knowledge, about the role of particularized, ethically charged experience in knowledge, and about the relationships between knowledge and political action. I have already noted that epistemic contextualism, as the dominant subliterature within traditional analytic epistemology focused on knowledge and context, is almost defined in ways that preclude a consideration of the nature of knowing agents that is requisite to taking up such issues. Providing an ironic manifestation of the idea that knowledge and interests are intertwined, there is subsequently very little knowledge about these dimensions to knowledge within traditional analytic epistemology, and seemingly little prospect for developing epistemic tools requisite for that knowledge.

Insofar as knowing agents have been explored within analytic epistemology more generally, they have been predominantly construed as universalized and idealized knowers, complete with some level of the kinds of properties that such knowers possess: rationality and reasoning, relevant empirical information, inferential knowledge, conceptual knowledge, and language. Knowing agents vary with respect to these dimensions of knowledge, but issues concerning the broader social contexts in which people come to know, their status as bearers of knowledge, and the relationship between their knowledge and their actions, all receive anemic treatment here. Epistemology is impoverished, at least with respect to the kinds of epistemic phenomena that these vignettes raise.

Apart from specific questions about knowledge in the context of intellectual disability, questions that I will return to in chapter 10, the vignettes raise more general issues about knowing agency from the margins that do not rely on that particular context. Historians, archivists, and others concerned with both the preservation of knowledge of the past and its active representation have devoted much thought and energy to the roles of survivor testimony, oral histories, and collective memory, especially when that past involves trauma, loss, and tragedy. Professionals in the medical and health sciences have begun, in the past ten years, to welcome more active roles for those on the receiving end of medical care—patients and parents and guardians of those patients—echoing a broader and older shift, reflected in the contemporary *citizen science movement*, in the relationship between scientific knowledge and practice and the public. In both the context of survivor knowledge and parental advocacy, there are two types of marginalization relevant to knowing agency from the margins.

First, both survivors and parents in general are marginal individuals themselves vis-à-vis the broader social systems of which each is a part. Survivors—of psychiatric care, of rape, of political persecution, of genocide, of residential schools, of sterilization—are outsiders to society because they have been institutionalized, because they have been sexually assaulted, because they have had family members killed or imprisoned, because they have been removed from their families in order to be "civilized," because they were deemed "unfit" for procreation and intelligent parenthood. This marginalization can be either antecedent or consequent to the very experiences conveyed by testimonies and oral histories. While parents are not marginalized agents in society in general, their role as agents acting in the interests of their children remains very much one as outsiders to the medical system, despite some acknowledgment of a place for parent-led advocacy in medical care. In both cases, survivor and parental agency is structurally compromised from the outset.

Second, both survivors and parents in general are marginalized as *knowing agents* vis-à-vis the systems for the production of knowledge about the relevant phenomenon. As illustrated by the conference report, survivor knowledge can be deemed less valuable than knowledge generated by professional historians, whether it be for putative shortcomings in authority, evidential robustness, and reliability, or for excesses in subjectivity and variability. Here the very positionality of survivors is the basis for challenges to taking their knowledge claims at face value. Likewise, being a parental advocate is viewed as occupying a compromised position from which to inquire, to issue directives that run counter to medical advice, or to question the authority and ethics of medical professionals. Despite being viewed by some as potential sources of knowledge and insight, in general parental advocates have a status as knowing agents that remains marginal and second-class.

A plausible hypothesis is that this second type of marginalization—as knowing agents—is a function of the first type of marginalization—as individuals. That is, the marginalizing outcome of one's location in the social order of things also makes those individuals second-class knowing agents in the systems that generate knowledge about that social order.

Exploring that hypothesis, however, is beyond the scope of this chapter and the next. What I am interested in pursuing here are strategies for challenging this latter form of marginalization, that of knowing agents. Indeed, I am interested in strategies that go further than challenging or even countering this marginalization—those that instead flip knowing agency from the margins on its head. Among such "inversion epistemologies," standpoint theory is the most sophisticated and best known.[5]

9.6   The Politics of Epistemic Apartheid

As I noted in section 9.2, the gendered disjuncture between the literatures on standpoint theory and epistemic contextualism itself raised questions about "Who knows?" and the politics of knowledge. Given the asymmetries of power and authority within epistemology in particular, and within philosophy more generally, the striking gender skew across these literatures warrants not only the attention of sociologists of knowledge but also of philosophers themselves. Not to care about the question "Who knows?" is not itself a crime; after all, it may just be "different strokes for different folks." But if those who care about the question "Who knows?" themselves remain knowing agents at the margins of the enterprise of understanding knowledge, then mere sociological variation becomes value-laden in ways that suggest the need for remedial measures beyond mere collective self-reflection. Strategies as simple as promoting more attention to knowing agency at the margins within traditional analytic epistemology, and facilitating bridging and integrative dialogues between epistemic traditions that treat the context of knowledge very differently, represent relatively low-cost ways to begin to address the political and ethical issues raised by mere sociological skew.

Although this point is directed primarily at traditional analytic epistemology and at philosophical approaches to understanding knowledge more
generally, in taking it up it is important to keep in mind the "standpoint
skew" within standpoint theory itself. Women may be knowing agents
from the margins of gendered oppression, but at least "feminist standpoint
epistemology" names a robust field within the study of knowledge, one
with its own rich and engaging literature. Indigenous and aboriginal people
are knowing agents from the margins of racialized oppression, and people
with a variety of disabilities are knowing agents from the margins of disability oppression. Analytic epistemologists might do more than merely
wonder about what is missing, epistemically speaking, in virtue of the near
complete absence of such knowing agents from the body of knowledge
that most forcefully occupies their own knowing agency and the epistemic
practices that it encompasses.

10   Eugenics Unbound: Survivorship for the Subhuman

10.1   Standpoint Theory and Knowing Agency

Having focused in chapter 9 on the politics of the marginalization of knowing agents at the margins within traditional analytic epistemology, I turn now to the question of *who knows* within standpoint theory itself. Although standpoint theory purports to provide a general account of knowing agents at the margins, as I noted in chapter 9, standpoint theory has developed with particular kinds of knowing agents in focus. As I shall suggest in this chapter, this complicates the application of standpoint theory in general, and standpoint epistemology more particularly, to a broader range of knowing agents, including those central to a standpoint eugenics. The centrality of survivors to the account of marginal knowing agents about both the eugenic past and the newgenic present that I have been advocating, together with the breadth of the conception of eugenic survivorship that I have employed, further complicate this application of standpoint theory to eugenics.

According to standpoint theory, despite their marginalized status, those who are systematically oppressed nonetheless possess some kind of epistemically privileged knowledge that both reflects their experience of oppression and plays a special transformative role in resisting the very structures of oppression that generate that experience. Thus, standpoint theory inverts the epistemic status of marginalized individuals vis-à-vis how they are regarded within traditional analytic epistemology, including within the literature on epistemic contextualism. There such marginalized knowers have no distinctive kind of epistemic status: they are simply universalized knowers who possess more limited access to epistemic resources, such as sources of information, the tools of knowledge, and positions of social authority. With those limitations of resource access, their knowledge is correspondingly restricted. And given that the potential to act, politically or otherwise, turns on that knowledge, such knowing agents also have a reduced capacity to effect social and political change via what they know.

Standpoint theory has always looked to bridge this distance between knowledge and action, stemming as it does from a longer tradition in which political commitment and values provide at times radical shifts in perspective. But as became clear in the debate around Susan Hekman's attempt to reinvigorate feminist standpoint theory, two dimensions to standpoint theory that are especially pertinent to underscoring how a kind of engaged, collective politics of knowledge is intrinsic to standpoint theory had gradually drifted out of focus.

First, the most important sort of knowledge generated by standpoint agents is value- and motivationally charged. That knowledge reveals something about how systemic oppression operates in particular cases; as such, it is intrinsically connected to action, including political action. Second, standpoint agents are effective as both knowers and agents not as isolated individuals, but as members of groups that have been, and often still are, subject to systemic oppression. This implies that the lived reality of the marginalized experience of standpoint agents has been built up historically, socially, and institutionally over time.[1]

10.2   Standpoint Agents for Class and Gender

As noted in chapter 9, standpoint theory in the discipline of philosophy has predominantly taken the form of feminist standpoint epistemology, which provides a framework for understanding how knowing agents subject to gender oppression occupy a standpoint affording distinctive kinds of epistemic insight into sexism and corresponding epistemic authority. In the original statements of standpoint theory, particularly by the sociologists Dorothy Smith and Patricia Hill Collins and the political scientist Nancy Hartsock, this approach to gender oppression was modeled explicitly on the account of class oppression and resistance found in classic Marxism. Just as the working class were knowing agents at the margins with respect to class oppression, so too were women knowing agents at the margins of gendered oppression. The structuring system governing each of these particular forms of oppression—respectively capitalism and patriarchy—operated through a dynamic opposition between pairs of agents—respectively, the ruling class and the working class, and men and women. And in operating oppressively, that dynamic opposition made those marginalized by that oppression special kinds of knowing agents, possessing what Collins calls an intimate "insider knowledge" about the ultimate workings of those structuring systems. Table 10.1 summarizes the putative parallel between these two original kinds of standpoint theory.

Capitalism and patriarchy, as the respective systems that generate and maintain class and gendered oppression, are constituted by institutions, practices, policies, and ideas, and the relationships among them. Standpoint theory contributes to the already abundant theorization of these systems and domains by claiming that, and articulating how, these structuring systems create certain kinds of knowing agents from the margins— standpoint agents.

A cluster of open issues within traditional standpoint theory revolves around the identification of standpoint agents and the nature of the knowledge that they possess. Perhaps the most central of these for articulating a standpoint eugenics is what we might call the *"Whose standpoint?"* question, originally raised by the philosopher Sandra Harding and much discussed over the past twenty-five years. This is the question of which knowing agents have or occupy standpoints, which provide those agents with epistemic privilege and authority of some kind. Since the working class and women were identified as such agents, respectively, for economic and gendered oppression in classic standpoint theory, the "Whose standpoint?" question calls for further interrogation of that view, minimally by way of some more nuanced explanation of precisely *why* the working class and women are such knowing agents. Thus, we might re-express the "Whose standpoint?" question as follows: in virtue of what do certain agents come to possess epistemically enhancing standpoints, despite the systematic oppression that they face?

Table 10.1

Standpoint theory for class and gender

Domain/Form of oppression	Structuring system	Standpoint agents
Economic status Class oppression	Capitalism	The working class
Gender Sexism	Patriarchy	Women

Although as table 10.1 indicates, it is in the first instance the working class who are oppressed under capitalism, and women who are oppressed under patriarchy, it seems that even classic standpoint theory is committed to the view that simply being working class or being a woman is not sufficient to possess the corresponding standpoint. This is because it is in virtue of their *situated experience* that the working class and women have insightful knowledge that escapes others, knowledge that itself is the starting point for resistance to the corresponding system of oppression. That situated experience, in turn, either encompasses or gives rise to a certain kind of awareness and process of group identification. Thus, there is an important place for consciousness-raising in the constitution of knowing agency. For this reason, occupying a standpoint has been construed as an *achievement* of some kind, one that requires a certain level of consciousness in a knowing agent, and this "achievement thesis," as it is sometimes called, has been emphasized within feminist standpoint epistemology. It follows from the achievement thesis that even if all women could be standpoint agents, only those with a certain kind of situated experience will actually become such agents.

The achievement thesis about who occupies particular standpoints has fueled what we might call *plus-or-minus* answers to the corresponding "Whose standpoint?" questions. Consider women as the default standpoint agents with respect to sexism. If only a subset of women—those with certain kinds of situated experiences—are standpoint agents with respect to sexism, as the achievement thesis implies, then we have a "women minus" answer to the "Whose standpoint?" question about sexism: some but not all women are standpoint agents here. Conversely, men with the experiences that constitute the relevant achievement should also share the privileged standpoint, generating a "women plus" answer to this question: women plus some men are standpoint agents here.

The insufficiency of mere class membership for knowing agency at the margins of capitalism was recognized in classic Marxist standpoint theory, and taking up the "Whose standpoint?" question here generates a parallel question. If it is a certain kind of situated experience, awareness, or group identification that makes for standpoint agency with respect to class oppression, can those beyond the working class be knowing agents at the margins? One of the issues correspondingly in play here is that of solidarity with those who directly experience gendered or class oppression.

In the historical development of Marxism, this issue played out primarily with respect to the role of intellectuals and other sympathetic members of the bourgeoisie in revolutionary change; with respect to gender, the issue is the place of men in feminism.

While plus-or-minus answers might seem simply an inevitable consequence of the nature of standpoint theory, I want to motivate another way of thinking of the "Whose standpoint?" question that opens space for other kinds of answer. That space should prove particularly welcome as we move beyond past focus within standpoint theory on class and gender. How should we think about knowing agency at the margins of race, sexuality, and disability, for example?

Standpoint theory *has* been applied to these domains, and those applications have yielded much insight about racial and disability oppression. Despite that, these extensions and applications of standpoint theory have not been explored as systematically as they have been in the cases of class and gender, and more particularly there has been little attention to some of the important dimensions of variation among different sorts of knowing agents. This inadvertent neglect has introduced some complacency about the ease with which at least classic standpoint theory can be extended to race, sexuality, and disability; it has perhaps also led to a blind spot regarding how those domains might, conversely, enrich standpoint theory itself.

I begin with a brief, partial, and necessarily selective consideration of the case of race, for two reasons. First, it is the example for which such an extension from class and gender is least problematic. Second, it is also the case that flags the kinds of challenges to and limitations of further extensions of classic standpoint theory to sexuality and disability in providing a putatively general account of knowing agency at the margins.[2]

10.3   Race, Lived Reality, and Group-Based Agency

The most developed work on standpoint theory with respect to race is that of Patricia Hill Collins, whose *Black Feminist Thought* has focused on the standpoint of, as she says, "US Black women" or simply "Black women." In concluding section 10.1, I noted that the knowledge possessed by standpoint agents was both value- and motivationally charged, and that the lived reality of the marginalized experience of standpoint agents has been built up historically, socially, and institutionally over time. Collins forcefully articulates a view of standpoint agents that is sensitive to both of these points by emphasizing that "no homogeneous Black *woman's* standpoint exists" and that "it may be more accurate to say that a Black *women's* collective standpoint does exist, one characterized by the tensions that accrue to different responses to common challenges." Collins's emphasis on the importance of a nonessentialist and collectivist understanding of standpoint agents is shared with that of other feminist standpoint theorists, and her focus on race has given that understanding a particular character.

As Collins emphasized in an exchange with Susan Hekman, people are born or enculturated into already-existing groups that often bear the experiential traces of a long history of systematic oppression. While some group membership and certain aspects of group membership may be more actively chosen by individuals, in at least many cases of race and ethnicity, group membership, like dependence in cases of developmental disability, is "thrown" rather than chosen. Indeed, a core part of the knowing agency of black women as situated knowers and agents lies in the political significance of the formation of group solidarity and community, a solidarity brokered not chiefly through consciousness-raising and other delicacies of middle-class feminism but via the lived history that group-based individuals experience. Anchored in the joint, lived reality of individuals as members of groups that have been constructed over time, the resulting solidarity belongs to individuals of a certain kind, with a shared history, culture, language, and way of being.

This rich, group-based conception of the relevant lived reality is important to Collins's approach to standpoint theory, but it also makes the "Whose standpoint?" question especially poignant. For although Collins herself moves freely between talk of "US Black women" and "Black women" as standpoint agents, the group-based lived reality here is at least partially shared with a range of others. In this regard, one might reflect on Collins's later response to the claim that "Black women are more oppressed than everyone else and therefore have the best standpoint from which to understand the mechanisms, processes and effects of oppression." After noting that "this is not the case," Collins continues:

Instead, those ideas that are validated as true by African-American women, African-American men, Latina lesbians, Asian-American women, Puerto Rican men, and other groups with distinctive standpoints, with each group using the epistemological approaches growing from its unique standpoint, become the most "objective" truths. Each group speaks from its own standpoint and shares its own partial, situated knowledge. But because each group perceives its own truth as partial, its knowledge is unfinished. Each group becomes better able to consider other groups' standpoints without relinquishing the uniqueness of its own standpoint or suppressing other groups' partial perspectives.

Although Collins's concern here is with the objectivity of knowledge generated from a multitude of standpoints, I am more interested in that multitude itself. The sort of pluralism about standpoints that Collins advocates is no doubt worthy of further discussion in its own right. The simple point here is that its mere existence significantly deepens the challenges posed by the "Whose standpoint?" question, especially as we move from class, gender, and race to disability.[3]

10.4   Generalizing Standpoint Theory

Recall that classic standpoint theory posits a number of structuring systems, each of which creates and maintains a corresponding form of oppression. For each such system, there is at least a prima facie kind of standpoint agent: the knowing agents at the margins for that particular domain of oppression. Consider then table 10.2, which provides a shorthand overview that expands the summary in table 10.1 in order to extend standpoint theory to race, sexuality, and disability.

Table 10.2
Standpoint theory: Adding race, sexuality, and disability

Domain/Form of oppression	Structuring system	Standpoint agents
Economic status Class oppression	Capitalism	The working class
Gender Sexism	Patriarchy	Women
Race Racism	White supremacy	People of color?
Sexuality Heterosexism	Compulsory heterosexuality	Homosexuals LGBTQ people?
Disability Ableism	?	People living with disability The subnormal?

There are two related differences between class and gender, on the one hand, and race, sexuality, and disability, on the other hand, relevant to the question of whether this simple expansion of standpoint theory is adequate to capture knowing agency at the margins in these domains. We will discuss them in turn, and at some length.

The first difference concerns whether oppression and marginalization are primarily oppositional or are instead what we might call *dissipative*. Sexism generated and maintained by a patriarchal system, and classism generated and sustained by a capitalist system, operate in an oppositional, binary manner with respect to opposed categories of agent: women and men, and the working class and the bourgeoisie. Moreover, male privilege turns on the oppression of women, and ruling-class privilege turns on working-class oppression. But as Collins's reflections on the variety of racialized standpoints suggests, it is less clear that racial oppression, as real as it is, operates in this manner, precisely because of the multitude of standpoints that exist in the case of racism. The problem that this signals in identifying the relevant standpoint agents is further exacerbated, I want to suggest, in the cases of both sexuality and disability, though I shall concentrate the following discussion on disability.

Note that it is the *oppositional* nature of these binary relationships, rather than simply the putative binary nature of the categories they involve, that creates the kind of marginalization central to standpoint theory. The key idea in classic standpoint theory, recall, is that the oppositional dynamic that constitutes the marginalization also creates knowing agents at the margins of knowledge. Thus, whether sex or gender themselves are *really* binary, or whether the division between the ruling class and the proletariat is exhaustive—important as they are in the context of broader theorizing about gender and class, respectively—are issues that are orthogonal to the claim being made here.

In contrast to the cases of sexism and classism, racism, heterosexism, and ableism do not appear to operate oppositionally in this way. Racialized, sexualized, and disability oppression and marginalization do not presuppose the kind of oppositional relationship between two (or more) kinds of agents that is at the heart of the account of knowledge agency at the margins provided by standpoint theory. Rather, that marginalization is promulgated by a kind of dissipative agency. A biological analogy may be the most evocative way to convey the contrast being drawn here.

In ecology, predator and prey stand in an oppositional binary relationship, one in which gains for one typically mean losses for the other, whether these concern adaptations, environmental circumstances, or rates of interaction. For example, the evolution of thicker or smoother skin by a prey animal (say a snake) to resist capture by its chief predator (say, a monkey) represents an adaptive gain for the prey over the predator, and for that very reason an adaptive loss for the predator in the evolutionary arms race the two are engaged in. Likewise, a predator that expands its night-hunting repertoire, and so increases its rate of interaction with its nocturnal prey brings about an ecological change in its favor, but one for that very reason tied to less fortunate outcomes for the prey. The oppositional relationship between predator and prey is the chief reason that predator-prey dynamics can be modeled by the Lotka-Volterra equations, which specify the rate of change in prey population as a result of changes in the predatory population (and vice versa), together with parameters describing their interactions.

By contrast, environmental damage or destruction does not stand in a similar oppositional relationship to some kind of agent, or result from such a relationship between two or more agents. Environmental damage or destruction can occur without living agents playing a key role at all—consider storms, earthquakes, and other so-called natural disasters—or through the actions of singular agents. It occurs often through the agency of human beings or through natural agency. But it is neither co-defined in terms of such agency nor dependent as a phenomenon on such agency. Rather, environmental damage or destruction can and does occur whether or not there are such agents.

In this respect, racialized, sexualized, and disability oppression are more like environmental damage than like the adaptive and ecological outcomes generated by the dynamics of predator-prey relationships. For these cases, while oppression has a negative effect, especially for certain individuals *qua* members of some particular group, that group oppression neither presupposes nor implies the existence of some other beneficiary group. The causes of the corresponding oppression or marginalization are dissipative.

The second difference concerns the role that an appeal to *systems* plays. In traditional standpoint theory for class and gender, this role is quite specific. Part of the account of knowing agency from the margins provided by Marxist standpoint theory turns on there being an economic system, capitalism, that structurally places two kinds of agents in opposition to one another. That idea was carried over to the domain of gender, where "patriarchy" names the corresponding structuring system with respect to women and men. Moreover, in each of these cases the very system creating class or gender oppression also provides for standpoint agency. Here the cases of race, sexuality, and disability differ.

We might wonder whether, for example, "white supremacy" and "compulsory heterosexuality" name structuring systems of oppression in the way in which capitalism (according to Marxism) and patriarchy (according to feminism) do. While they may be, and often have been, pressed into service to do so, they are more typically used to refer to the phenomenon of racism or heteronormativity themselves, and as such play the role of *explanandum* rather than *explanans* in theoretical accounts of race and sexuality. In the case of disability, there is not even a corresponding such term; inventing the term "compulsory able-bodiedness" to play this role would only serve to underscore the distinctness, in this respect, of ableism from class oppression and sexism.

These two considerations should provide at least some pause to those assuming that standpoint theory can be generalized in a relatively straightforward way to all cases of knowing agency from the margins, or that one can simply presume its suitability for all such cases. Viewing race, sexuality, and disability through the classic standpoint lens that focuses our understanding of class and gender may even be counter-productive or limiting for advancing the very idea of an epistemology for other kinds of knowing agency from the margins.

I turn next to make some constructive suggestions for reimagining knowing agency from the margins by concentrating on intellectual disability, and so returning more squarely to the knowing agency most central to a standpoint eugenics. Doing so, I shall suggest, introduces some possibilities for standpoint epistemology more generally that involve moving beyond what I have called plus-or-minus answers to the "Whose standpoint?" question.

10.5   Joint and Extended Action in Intellectual Disability

At least some people with disabilities seem prima facie unlikely to be standpoint agents. People with *intellectual* disabilities in particular are often presumed not to have, and sometimes in fact may not have, the requisite cognitive capacities to be the sorts of knowing agents whose experience constitutes the basis for some kind of radical epistemic upheaval. Consider just that subset of people with intellectual disabilities who have "severe cognitive impairments." Such people have capacities that limit the extent to which they can share or participate in the very sorts of things—perspectives, narratives, community, consciousness-raising—at the heart of standpoint theory. Indeed, given a plus-or-minus answer to the "Whose standpoint?" question, people with severe cognitive impairments provide a reason to think that not all people with disabilities are standpoint agents. Despite the appearances here, I want to suggest a reconceptualization of what knowing agency involves that sees all people with disabilities as potential standpoint agents.

The kinds of dependency that result from severe cognitive limitations are recognized legally through the provision of legal guardianship, a relationship similar to that between parent and child. For this reason, the guardian is typically taken to be the knowing agent, where he or she acts on behalf of, and in the best interests of, his or her ward. There are, however, two alternative ways to conceptualize knowing agency here that are both more informative about the kind of knowing agency that people with intellectual disabilities can possess, and instructive for reconceptualizing the "Whose standpoint?" question.

First, the knowing agency of people with intellectual disabilities could be conceptualized as a type of *joint or shared agency*, whereby (for example) a person-guardian pairing acts jointly, or a person with an intellectual disability and her care provider jointly explore some object or opportunity of interest. Accounts of joint agency have been developed independently within action theory and the philosophy of science and cognitive science. At least some forms of such agency presuppose the kind of *collective intentionality* on the part of the co-participants in the action, a sort of sharing of knowledge that is distributed across the dyadic group that I have already discussed in chapter 6 as part of our distinctive human sociality.

Second, this knowing agency could be conceived as a form of *extended agency*, a notion modeled on extended *cognition* familiar within the philosophy of mind and cognitive science and that I have drawn on in part II in developing the socio-cognitive framework for understanding the puzzle of marked variation. Here we might view the guardian or care provider as an agential extension of the person for whom he or she is a guardian or care provider. Hence, we would view guardians or care providers functionally, so to speak, in the same way that we view cognitive technologies—screen readers, cellphones, pen and paper. Like such technologies that enhance our cognition and the action that it drives, such people function to provide extensions to an individual's agency. While such a view of agency does not itself presuppose collective intentionality, it does mark a potentially radical departure from historically influential individualistic views of both cognition and agency, as well as autonomy.

A modified model of knowing agency that incorporates people with intellectual disabilities, including those with severe cognitive impairments, as knowing agents via the notions of joint and extended agency, provides an alternative to a kind of hyper-individualistic strand to traditional standpoint theory, one that it shares with traditional analytic epistemology. It does so by reminding us that knowing agency itself is not always, and perhaps less often than we think, a matter of the situatedness of isolated and autonomous individuals. Rather, as we saw in the case of race and the sense in which standpoint agency is group-based, there is often a "thrownness" to knowing agency that lies in cooperation among others, the dependence of standpoint agents on others, and their shared agency with others. That is what their situated experience *is*, and we need conceptions of knowing agency at the margins that are sufficiently broad to encompass the kinds of agential entwinement that this entails. Thus, contrary to plus-or-minus answers to the "Whose standpoint?" question, the knowing agency of standpoint agents at least sometimes involves joint and extended agency. It manifests solidarity based on lived experiences anchored in a shared, group history, building a deeper sense of community and *mitgefuhl* (compassion for, or feeling sympathy with) through that socially located solidarity.

We might go further and challenge an additional assumption underlying plus-or-minus answers to the "Whose standpoint?" question. This is the reifying assumption that standpoint agents are always *sorts of people*, such as women or working-class people. Consider again intellectual disability.[4]

10.6   Intrinsic Heterogeneity and Sorts of People

The boundaries of the category of "intellectual disability" are contentious in various ways. The category may include people with autism, people with one or more of a variety of speech disorders, people with one of a number of trisomies (the best known, as we saw in chapter 7, is Down syndrome), older people with Alzheimer's or Huntington's disease, and people who have incurred brain injuries resulting in cognitive limitations. Even if the proper characterization of intellectual disability excludes some of these, the problem of heterogeneity that I am pointing to here remains and arises for related categories, such as "psychological disability." In addition to the preceding sorts of people, this category also includes people who have suffered trauma sufficient to cause post-traumatic stress disorder or borderline personality disorder, people with anxiety disorders or very specific phobias (e.g., agoraphobia or arachnophobia), people with other disorders of emotion or personality, such as bipolar personality disorder or narcissism, schizophrenics and others suffering from visual or auditory delusions, or delusions of self-perception (e.g., self-aggrandization). The significant differences between people across these categories in terms of social location, functionality, and life prospects itself calls into question the value of clustering them together under the heading of "psychological disability," particularly if they are taken to occupy a standpoint in virtue of belonging to this group of people.

To put it bluntly, "intellectual disability," like "the feeble-minded" or "the mentally deficient" before it, is a reification. As we saw in part I, the feeble-minded and the mentally deficient were reified by eugenics—as applied science and as social movement—and that reification played a key role in the subhumanization enacted in the social mechanics of eugenics. The category "intellectual disability" is not only a descendant of these past reifications, but like them glosses over important differences between those whom it categorizes. Because of this, there are no standpoint agents fitting the description of "people with intellectual disabilities." This is not because there are no distinctive, situated, individual experiences here; it is because they are not the experiences of a distinctive sort of agent.

While the heterogeneity in the case of intellectual and psychological disability is partially a function of the medicalization of disability more generally and its history, the heterogeneity I am pointing to also exists in the less heavily medicalized cases of race and sexuality. Those undergoing racialized oppression may be African Americans living in the wake of Southern slavery, Roma people who were sterilized in Baltic and Eastern European countries in the second half of the twentieth century, or First Nation children who were taken from their parents and sent to residential schools until 1996 as part of a government-sponsored "civilizing mission" to "breed the Indian out of the child" in Canada. Sexualized oppression is the oppression not only of people who have a particular sexual orientation, such as homosexuality or bisexuality, but also of people who are transsexual or transgender, people with nondominant sexual preferences, and even people with certain kinds of views of sexual practices and activity.

Again, the intrinsic heterogeneity here does not make the oppression any less real or impactful. What it calls into question, rather, is the adequacy of conceptualizing the oppression as operating on particular *sorts of people*, for those categories are themselves reifications. Standpoint agents subject to racism, heterosexism, and ableism—unlike those subject to sexism and classism—are not sorts or kinds of people at all, as standard construals of the "Whose standpoint?" question presuppose. Rather, at best they are disjunctive kinds, and at worst they are simply defined in terms of what they are not: not white, not straight, not normal.[5]

10.7   Recovering the Voices of Eugenic Survivorship

Central to an adequate account of knowing agency from the margins—whether it be a form of standpoint theory or merely standpoint-ish eugenics—are the voices of eugenics survivors in the broad sense in which I have used that expression. The marginalization of eugenics survivors is in part the silencing of their voices, of the narratives they produce as knowing agents with a special kind of knowledge of eugenics. Here standpoint eugenics has a key role to play in the recovery of these voices.

To return to the discussion of silencing in chapter 7, standpoint eugenics shifts the discursive authority to those who have been targeted by eugenic logic and edged out of sites of knowledge production. Contesting and reclaiming these sites is to recognize that self-advocates and those otherwise targeted by eugenic logic are the authorities of their oppression and must lead the way in disability politics. It is to recognize that the onus of proving "harm" in debates of liberal reproductive ethics does not take place on a neutral playing field, but is skewed by eugenic logic. The perspectives of actual and putatively disabled people must have greater weight in these discussions.

The recovery of voice in the context of disability oppression, however, is rarely a matter of sitting back and simply being receptive to the production of vocal noises emanating from the survivors of eugenics. As in other such contexts, active effort is needed to create the conditions in which that recovery can take place. A core part of the work that we undertook in the Living Archives project was to create the context in which the stories of eugenics survivors could be told, both as individual stories owned in a deep sense by those who tell them, and collectively as an exercise in community building and bonding between survivors themselves and in bridging between the nascent community that that bonding creates and the local communities from which survivors had been excluded through stigmatization, discrimination, and voicelessness.

As touched on earlier, Leilani Muir's legal case against the Province of Alberta and the cases that were settled out of court in its wake revealed much that, previously, had been virtually unknown about the history of eugenics in Western Canada. This includes aspects of day-to-day life for those residing in institutions such as the Provincial Training School of Alberta in Red Deer. It also includes facts about the treatment of children in these institutions recounted in chapter 4—treatment that many continue to find deeply disturbing. Paramount among these facts was the use of bare "quiet rooms" in which residents could be left in relative isolation for days or weeks and, during this time, the apparently extensive psychotropic experimentation on children deemed mentally deficient or psychotic, and the revelation of a number of cases in Alberta during the 1950s and 1960s in which testicular tissue was unnecessarily removed during eugenic castrations (not simply sterilizations via vasectomies) performed on males with Down syndrome, tissue that was then used for experimental research purposes. Part of what is truly disturbing here is that males with Down syndrome are already reproductively sterile, and they were known to be during this time. Another part of what is so disturbing lies in the discretion of the Eugenics Board to, in effect, grant permission for tissue removal for medical experimentation, clearly exceeding the authority it had been given by the Sexual Sterilization Act of Alberta.

Like other oral histories, those of people institutionalized, sterilized, and stigmatized by Alberta's eugenic enthusiasm convey something both of what life was like in the past—in their case, in government-run institutions such as the Provincial Training School—and of what life is like having survived a traumatic past—in this case, a eugenic past. Apart from the distinctive content provided by survivor standpoints, the resulting narratives are notable for being coherent, partial, and personally reflective stories from those thought incapable of this kind of narration. Here the restoration of full humanity that narrative brings counters both the dehumanization caused by traumatic episodes and events and the subhumanization of people whose putative reasoning deficits excluded them from an important part of family life and was the basis for their extended segregation from the rest of society. That restoration also helps to build some sense of community feeling among survivors, both with one another and with others participating in a common project. Since today's popular views of cognitive disability share much with those of the middle half of the twentieth century, people who have, or are assumed to have, cognitive disabilities continue to confront subhumanizing attitudes and policies. For these reasons, the kind of oral history developed by a standpoint eugenics has a special role to play in the collective story we tell about eugenics and disability in Canada, and perhaps more generally.[6]

10.8   Narratives, Stories, and Standpoint

The voices and narratives of eugenics survivors themselves, as preserved at the Our Stories module of the Living Archives website, are the project's most valuable resource. Here I provide a sample of those voices as manifest in three of the stories that have been constructed. The first two are stories told by sterilization survivors who are now in their sixties; the third is a story told by a much younger person, and it reflects on his own experiences with disability and parenting.

Early in his story, Roy Skoreyko tells us that he's now retired and enjoys what he does, adding, with a laugh, that he "has a life." He then reflects on some of his experiences at the Provincial Training School, now called the Michener Centre:

The first day that I moved into Michener Centre, I was about ten years old. My parents dropped me off there, and man it was so hard for me to leave my family. All they did was give me some other clothes. I couldn't wear my regular clothes; they gave me jeans and all that, and a shirt. And they would just drag you and put you in your day room there. [...] And we all had to line up to the dining room to eat, lined up to go take a shower. When you lined up to take a shower, you had to walk down the hallway, no clothes on at all, and people could see you. [...] It was scary, 'cause the doors were always locked. And they just went like this [moves hands together] [...] and they always had a key, to lock the dayroom doors, lock the dormitory doors where people would go to sleep [...] Some of us were scared to say anything, scared for our lives. And I have seen some of the people there that were tooken [sic] into this little room by themselves. No clothes on, and all there was just [...] there was no sink, all there was was just a mattress on the floor, and there was no bed, no bed at all. You had no clothes on. And you could see, they had these thick, thick glass windows, but you still could see people walking around in there. And you would be locked in there.

Beginning with his feelings about being "dropped off," Roy moves quickly to say how clearly he remembers the regimentation of life at Michener—how he was clothed, the lining up, the locked doors, the "quiet rooms," his nakedness, being seen by others. Returning later in his story to talk about his ongoing, positive relationship with his parents while he was at Michener—including visits home, his reluctance to return to Michener, and his discussions with them later in life about his sterilization—Roy ends his story by reflecting on disability and his pride in how far society has come. Of his sterilization, Roy simply says: "I feel offence and a little bit disappointed that that was done. You know, they took control of our body, and they did it, and they didn't ask us. There was a lot of things that went on there that, you know, shouldn't have."

The second story is that of Glenn George Sinclair. Glenn, who is soft-spoken, tells his story by sharing photograph albums and material from his legal case file. For Glenn, as for other sterilization survivors, while the photograph album prompts many personal memories, the legal file contains information (from institutional and other government authorities) about his life that, for many years, had remained unknown to him. Glenn has spent much time reading and reflecting on this "information." Glenn describes both how he felt living in the Provincial Training School and what he took to be the societal attitudes reflected in his treatment there:

I felt kinda like I was just gonna be there and live like, live my life, like a zombie you know, like lock 'em up, throw the key away. Like I didn't really have any purpose in life. You're commanded to do things, they yell at you, "do this, do that," and you can't do this or you can't do that. You felt like you were being ordered around like a dog, like an animal in a cage sort of thing, you know. You didn't feel human at all. You just feel as if you, as if you exist, you know. Like you feel nothing, like there's no hope.

Despite their brevity, Glenn's comments on the documents contained in the legal file are powerful, especially when juxtaposed with their visual presence in the video story. Part of what Glenn, Roy, and other survivors grapple with are the views expressed about them in these documents and, in turn, the broader societal attitudes toward the "mentally deficient" that they convey. Glenn does not need to say much in order to convey the absurdity and wrongfulness of his labeling and subsequent subhumanizing. His visual presence in telling his story provides an immediate, direct, and rehumanizing connection to anyone listening to and watching him.

The third story is not that of a sterilization survivor but, rather, someone more than a generation younger than Roy and Glenn—someone whose reflections resonate with much of what survivors have to say about family life and disability. Kyle Lillo communicates nonverbally, and his experience with a life-changing acquired disability in Alberta positions him as a eugenics survivor in the broad sense introduced in chapter 1. Kyle's story is presented via photographs and music, and he uses text-to-voice software to present a verbal narrative. Kyle's story focuses on his positive views of his family and friends, his work with children at the Glenrose Hospital, his status as a role model for children (especially given his own history), and his views of the rights and responsibilities of parenting. After saying, "I know that I am not ready to start a family right now," Kyle notes what a big commitment this would be. He then turns to the views of others:

I am sure that some people would react poorly if I became a parent. But those people do not know the real me, so their opinions do not matter. I have every right to be a parent. I am a good person who knows how to love. No one can tell me otherwise and I would fight for the right if somebody told me I couldn't have children.

Kyle then lists some of the things that he would help with, and some that he would not need help with, once he is ready to start a family. Kyle's story concludes with a short reflection on government intervention and an affirmation of his right to parent: "I don't think that the government would let me keep my child. But like I said before, I would fight for my rights to make sure I keep my child. The government has no right to tell me that I am not fit to be a parent."

Like the stories of Roy and Glenn, Kyle's story focuses on what he thinks is important to express both about himself and his views. In constructing these stories, each storyteller has authorial control through active participation in the extensive editing and reediting of his story as well as the opportunity to give, give again, and retract consent to the story that evolves throughout the process.

Although all three stories are told in distinctive ways and convey much about the unique personalities of each of the storytellers—their engaged individuality—as I hope even this brief look at them suggests, they overlap and contrast in rich and provocative ways. For example, the conclusion to Roy's story shares a strong sense of disability activism with Kyle's story, while Glenn's reflections on how he was treated convey more of a sense of a personal injury, one that was, as Glenn acknowledges, shared with many others but is nonetheless very much his own. Glenn's and Roy's stories both bring out some of the fears, humiliation, and sense of day-to-day trepidation that were part of institutional life, whether brought about by being paraded through hallways naked (as in Roy's case) or by feeling that the institution's attitude toward them was "lock 'em up, throw the key away" (as in Glenn's case).[7]

10.9   Concluding Thoughts

I began part III by contrasting the disinterestedness of traditional analytic epistemology in questions about "Who knows?" and about the politics of knowledge more generally, with standpoint epistemology's focus on the subjects of knowledge and the variation between those who know. Standpoint approaches to class and gender have provided systematic accounts of the knowledge possessed by certain kinds of standpoint agents, such as women and the working class, and in this chapter I have concentrated on questioning how readily those insights about standpoint agency could be generalized.

A point familiar from feminist standpoint epistemology is that the community that philosophers in particular have built for themselves has excluded certain kinds of voices. Those marginal voices include indigenous people and people with intellectual disabilities, people whose near complete absence from the work and awareness of professional philosophical community perpetuates the silencing of their voices in the broader community. Recognizing the engaged individuality that eugenics survivors have requires more actively rethinking both the spaces and places in which philosophy is undertaken, as well as the conceptual tools that are used in those undertakings.

In exploring the phenomenon of knowing agency from the margins beyond the cases of gender and class, I have tried to identify some of the resources available for a more complete development of standpoint theory, one that accords a place for standpoint eugenics. In section 10.3, I focused on the case of race in order to bring out two points. The first is that the group-based nature of knowing agency at the margins is often narratively anchored in shared histories of oppression rather than in newly-recognized group identities. The second is that emphasizing a richer notion of lived experience that appeals to the thrownness of group-based knowing agency complicates the answer to the "Whose standpoint?" question, making it more difficult to identify "the" standpoint agents for race. In section 10.4, I argued that racism, heteronormativity and ableism do not function oppositionally but dissipatively and are not governed by an overall structuring system, as classism and sexism are presumed to be in classic standpoint theory. Thus, we need to more critically interrogate the assumption of classic standpoint theory that we can simply generalize it to all standpoint agents. On the view of knowing agency at the margins that derives from a focus on intellectual disability, knowing agency is sometimes epistemically shared, distributed, and cognitively extended (section 10.5), and involves groups that are intrinsically heterogeneous and only problematically thought of as designating sorts or kinds of people at all (section 10.6). Whether a true standpoint eugenics can accommodate both of these points about intellectual disability, or whether they push one to move beyond the resources of standpoint theory, remain issues that I have not tried to resolve here.

While acknowledgment of both of these features of knowing agency from the margins can be found within pockets of standpoint theory, it is no accident that they are features to be found most prominently in the most marginal cases of knowing agency from the margins: those involving intellectual disability. Standpoint theory itself has not been exempt from the dynamics about which it has theorized. Developed primarily from the standpoint of women as part of a feminist epistemology, the knowing agency from which it has drawn its insights does not equally reflect the standpoint of all standpoint agents. How could it?

How adequately standpoint theory can encompass the full diversity there is to knowing agency at the margins will turn very much, in my view, on how well it can accommodate the kinds of features of knowing agency summarized earlier. A radicalization of Collins's pluralistic standpoint theory for race that advocates as many standpoints as there are sorts of people problematizes the "Whose standpoint?" question. Likewise, modifying standpoint theory to embrace joint standpoints, or standpoints with *no* sorts of people, may push one beyond the boundaries of standpoint theory as an account of knowing agency from the margins.

In any case, the oppositional and system-based account of the generation of standpoint knowledge requires further scrutiny from those seeking a fully general account of knowing agency from the margins. Creating space in that account for the unique perspectives that eugenic survivors provide, not only on the history of eugenics but also for our collective reflections on eugenics, newgenics, and disability, is one initial, constructive step beyond such scrutiny. Challenging the ongoing marginalization of such accounts of knowing agency from the margins is another.

One final note on the centrality of the idea of sorts of people to my account of The Eugenic Mind. In the Galtonian view of eugenics developed in chapters 1 and 2, we saw that it was the differential value assigned to various sorts of people, and thus their putative reproductive potential for intergenerational human improvement, that marked off eugenics from other social interventions aimed at bettering our species. In chapters 3 and 4 I argued that eugenic traits in effect defined sorts of people who were subhumanized in the eugenic past as well as in contemporary discussions of cognitive disability within rationality-based approaches to ethics. And in chapters 5 and 6 I introduced the puzzle of marked variation as well as the socio-cognitive framework as a response to that puzzle in terms of there being variant sorts of people to which we differentially respond. In chapter 7 I discussed how the practice of prenatal testing coupled with selective abortion (and in chapter 8, the wrongfulness of eugenic classification, institutionalization, and sterilization) involved, in their own way, the mistaken reification and application of categories of disability, particularly intellectual disability.

Thus, fairly resolute skepticism about whether there really are the sorts of people that stock The Eugenic Mind has structured much of my discussion in parts I and II. It seems ironic and fitting to conclude part III's more metatheoretical discussion of the adequacy of standpoint theory to account for knowing agency at the margins in the case of eugenics and disability by returning, in part, to that same skepticism. My hope is that standpoint eugenics can be further developed with no vestige of The Eugenic Mind itself.

Notes

Author's note: Departing from MIT Press house style, these notes are summative and self-contained, with at most one note per section directing readers both to strict citations and to relevant readings for that section of the chapter. The aim here is to enhance the flow to the text as well as the utility of the notes themselves for those wishing to pursue some of the issues further. The consistency of this style of notation is disrupted just twice, in each case when that joint aim is better served by including brief page references in the text as part of a more detailed examination of a given, influential view: in section 5.5 in discussing Lennard Davis's views of normalcy and disability, and in section 7.5 in taking up Julian Savulescu and Guy Kahane's defense of the principle of procreative beneficence.

1 Standpointing Eugenics

1. Feminist standpoint theory was first articulated in the work of Dorothy Smith, "Women's Perspective as a Radical Critique of Sociology," *Sociological Inquiry* 44 (1974): 7–13; Nancy Hartsock, "The Feminist Standpoint: Developing the Ground for a Specifically Feminist Historical Materialism," in *Discovering Reality: Feminist Perspectives on Epistemology, Metaphysics, Methodology, and the Philosophy of Science*, ed. Sandra Harding and Merrill Hintikka (Dordrecht: D. Reidel, 1983), 283–310; and Patricia Hill Collins, "Learning from the Outsider Within: The Sociological Significance of Black Feminist Thought," *Social Problems* 33, no. 6 (1986): S14–S32. For collections of influential essays on feminist standpoint theory, see Sandra Harding, ed., *The Feminist Standpoint Theory Reader: Intellectual and Political Controversies* (New York: Routledge, 2004); a special issue of *Signs: Journal of Women and Culture* 22 (1997); and a collection of reflective papers in *Hypatia: A Journal of Feminist Philosophy* 24, no. 4 (2009): 189–237. The most sustained and informative articulation of standpoint theory with respect to class in the earlier Marxist tradition is Georg Lukács, *History and Class Consciousness: Studies in Marxist Dialectics* (Cambridge, MA: MIT Press, [1923] 1971). For some brief discussion of feminist standpoint theory and disability, see Susan Wendell, *The Rejected Body: Feminist Philosophical Reflections on Disability* (New York: Routledge, 1996): 69–74, and for more extensive

discussion, Mary B. Mahowald, "A Feminist Standpoint," in *Disability, Difference, Discrimination: Perspectives on Justice in Bioethics and Public Policy*, ed. Anita Silvers, David T. Wasserman, and Mary B. Mahowald (Lanham, MD: Rowman & Littlefield, 1998), 209–252.

2. The full title of that most famous work of Charles Darwin is *On the Origin of Species by Means of Natural Selection, or the Preservation of Favoured Races in the Struggle for Life*, first published in 1859. Francis Galton's earliest work on eugenics is his two-part "Hereditary Talent and Character," *Macmillan's Magazine* 12 (1865): 157–166, 318–327, which formed the basis for his first book in the general area, *Hereditary Genius: An Inquiry into Its Laws and Consequences* (London: Macmillan, 1869). Galton coined the term "eugenics" in his *Inquiries into Human Faculty and Its Development* (London: Macmillan, 1883).

3. For the most influential view of how disability was regulated by a concept of normalcy constructed within eugenics, see the disability studies scholar Lennard Davis's *Enforcing Normalcy: Deafness, Disability and the Body* (London: Verso, 1995). For earlier constructivist work in a similar fashion focused on the disciplinary location of eugenics, see the sociologist Nikolas Rose's *The Psychological Complex: Psychology, Politics and Society in England 1969–1939* (London: Routledge and Kegan Paul, 1985). For recent work on prosociality, see David A. Schroeder and William G. Grazziano, eds., *The Oxford Handbook of Prosociality* (New York: Oxford University Press, 2015).

4. The quotation from Kevles's book *In the Name of Eugenics* provided at the beginning of this section is from page ix of the 1995 edition. Work by historians on North American eugenics traces back to Mark Haller's *Hereditarian Attitudes in American Thought* (New Brunswick: Rutgers University Press, 1963) and much of the more general academic work has continued with a focus on the United States, including Kenneth M. Ludmerer's history of genetics, *Genetics and American Society* (Baltimore, MD: Johns Hopkins University Press, 1972), which devotes much attention to eugenics. Daniel Kevles's *In the Name of Eugenics* (Cambridge, MA: Harvard University Press, [1985] 1995), which draws its title from an interwar quotation from the eminent British biologist J. B. S. Haldane, remains the authoritative work on science and eugenics, and covers Great Britain as well as the United States. Troy Duster's *Backdoor to Eugenics* (New York: Routledge, [1990] 2003) remains a much-discussed sociological perspective on eugenics past and present. For overviews of what is now a large literature, see the bibliography assembled by Paul Lombardo and Greg Dorr in 2000, accessed October 1, 2016, http://buckvbell.com/othermaterial.html; the historiographical review by David Cullen, "Back to the Future: Eugenics—A Bibliographic Essay," *The Public Historian* 29, no. 3 (2007): 163–175; and my own recent, annotated bibliography, "Eugenics and Philosophy," *Oxford Bibliographies Online*, accessed August 1, 2017, http://www.oxfordbibliographies.com/. For critical discussion of the Human Genome Project, see Daniel Kevles and Leroy Hood, eds., *The Code of Codes: Scientific and Social Issues in the Human Genome Project* (Cambridge,

MA: Harvard University Press, 1992), and papers by Marga Vicedo, Philip Kitcher, Diane B. Paul, Elisabeth A. Lloyd, and Alexander Rosenberg collected in part VIII of David L. Hull and Michael Ruse's anthology *The Philosophy of Biology* (New York: Oxford University Press, 1998). For reflection on the long-reach of Scandinavian eugenics, which like that in Alberta continued into the 1970s, see Gunnar Broberg and Nils Roll-Hansen, eds., *Eugenics and the Welfare State: Sterilization Policy in Denmark, Sweden, Norway, and Finland* (East Lansing: Michigan State University Press, 1996) and "Eugenics in Scandinavia," special issue, *Scandinavian Journal of History* 24, no. 2 (1999), ed. Gunnar Broberg and Mathias Tyden. The Broberg and Roll-Hansen volume was published in a revised edition in 2005 in part due to the developments in North America that I discuss here. For the disability rights critique of selective abortion, see Adrienne Asch, "Disability Equality and Prenatal Testing: Contradictory or Compatible?," *Florida State University Law Review* 30, no. 2 (2003): 315–342, and her "Why I Haven't Changed My Mind About Prenatal Diagnosis: Reflections and Reminders," in *Prenatal Testing and Disability Rights*, ed. Erik Parens and Adrienne Asch (Washington, DC: Georgetown University Press, 2000), 234–258. For the contemporary resonance of the eugenic past in academic scholarship, see Alison Bashford and Philippa Levine, eds., *The Oxford Handbook of the History of Eugenics* (New York: Oxford University Press, 2010); the special issues of the *International Journal of Disability, Community, and Rehabilitation* 12, no. 2 (2013), ed. Gregor Wolbring, and the *Canadian Bulletin of Medical History* 31, no. 1 (2014), ed. Erika Dyck; books on North American eugenics, such as Paul Lombardo's collection, *A Century of Eugenics in America: From the Indiana Experiment to the Human Genome Era* (Bloomington: Indiana University Press, 2011) and Randall Hansen and Desmond King's *Sterilized by the State: Eugenics, Race, and the Population Scare in Twentieth-Century North America* (New York: Cambridge University Press, 2013); books focused on Alberta in particular, such as Jane Harris-Zsovan's *Eugenics and the Firewall: Canada's Nasty Little Secret* (Winnipeg, MB: J. Gordon Shillingford Publishing, 2010) and Erika Dyck's *Facing Eugenics: Reproduction, Sterilization, and the Politics of Choice* (Toronto: University of Toronto Press, 2013); eugenics survivor testimony and reporting based on such testimony, such as Kevin Begos, Danielle Deaver, John Railey and Scott Sexton, *Against Their Will: North Carolina's Sterilization Program and the Campaign for Reparations* (Florida: Gray Oak Books, 2012) and Leilani Muir, *A Whisper Past: Childless after Eugenic Sterilization in Alberta* (Victoria, BC: Friesen Press, 2014); and the documentary film, *Surviving Eugenics* (Vancouver, BC: Moving Images Distribution, 2015), co-directed by Jordan Miller, Nicola Fairbrother, and Robert A. Wilson.

5. For the landmark legal case, see Muir v. Alberta, *Dominion Law Reports* 132 (4th series) (1996), 695–762, 1996 CanLII 7287 (QB AB); and the film *The Sterilization of Leilani Muir* (Ottawa: National Film Board of Canada, 1996). For easy access to introductions to Canadian eugenics province-by-province, see the EugenicsArchive.ca provincial map, accessed October 1, 2016, http://eugenicsarchive.ca/discover/world. For the standard work on Canadian eugenics written before the Muir case, see Angus

McLaren, *Our Own Master Race: Eugenics in Canada, 1885–1945* (Toronto: McClelland & Stewart, 1990). For more recent work that focuses on Alberta, see Jana Grekul, Harvey Krahn, and David Odynak, "Sterilizing the 'Feeble-Minded': Eugenics in Alberta, Canada, 1929–1972," *Journal of Historical Sociology* 17 (4) (2004): 358–384, and Erika Dyck, *Facing Eugenics: Reproduction, Sterilization and the Politics of Choice* (Toronto: University of Toronto Press, 2013). The psychologist Henry Goddard introduced the term "moron" in a 1910 report, and it can be found in his better known *The Kallikak Family: A Study in the Heredity of Feeble-Mindedness* (New York: Macmillan, 1912) and *Feeble-Mindedness: Its Causes and Its Consequences* (New York: Macmillan, 1914).

6. Among MacEachran's few publications are a pair of general papers published in 1932, shortly after he was appointed as the chair of the Eugenics Board: John MacEachran, "A Philosopher Looks at Mental Hygiene," *Mental Hygiene* 16 (1932): 101–119; and "Crime and Punishment," address to the United Farm Women's Association of Alberta, reprinted in *The Press Bulletin* 17 (6) (1932): 1–4. For the Muir case, see the preceding note; the quotation from Madame Justice Veit comes from paragraph 3 on page 695. For Bill 26, see the entry on it (including full text) at EugenicsArchives.ca, accessed October 1, 2016, http://eugenicsarchive.ca/discover/connections/5233cd305c2ec500000000a5.

7. For the MacEachran Report, see "Report of the MacEachran Subcommittee, Department of Philosophy, April 1998," submitted by David Kahane, David Sharp, and Martin Tweedale. Until recently, this was available from the University of Alberta's Department of Philosophy website, but now see https://s3.amazonaws.com/bmcmahen/maceachran_report.pdf, accessed July 26, 2017. The psychologist Douglas Wahlsten also played a key role following the Muir case in drawing broader attention to MacEachran's eugenic efforts; see, for example, his "Leilani Muir versus the Philosopher King: Eugenics on Trial in Alberta," *Genetica* 99 (1997): 185–198.

8. For Leilani Muir's life, see her autobiography, *A Whisper Past: Childless after Eugenic Sterilization in Alberta* (Victoria, BC: Friesen Press, 2014), and her video narrative at the Our Stories module at EugenicsArchives.ca, accessed October 1, 2016, http://eugenicsarchive.ca/discover/our-stories/leilani. Stories from other eugenics survivors can also be found there, including those for Judy Lytton and Glenn George Sinclair. For recent work on dehumanization, see the philosopher David Livingstone Smith's *Less Than Human: Why We Demean, Enslave, and Exterminate Others* (New York: St. Martin's Press, 2011), his "Dehumanization, Essentialism, and Moral Psychology," *Philosophy Compass* 9 (2014): 814–824, and his short entry, "Dehumanization: Psychological Aspects," at EugenicsArchives.ca, as well as the comprehensive psychological review provided by Nick Haslam and Stephen Loughnan, "Dehumanization and Infrahumanization," *Annual Review of Psychology* 65 (2014): 399–423.

9. The Community-University Research Alliance Program of the federal funding agency The Social Sciences and Humanities Research Council of Canada was created in 2000 and closed in 2013, being replaced by two funding opportunities within SSHRC's newly-created Connection Program. There were ten CURA grants of $1 million awarded annually, with each project running for five years. The Living Archives on Eugenics in Western Canada project (www.eugenicsarchive.ca) was awarded funding in February 2010, and was completed in August 2015 after an extension was granted for completion. The video narratives referred to can be found at the Our Stories module at EugenicsArchive.ca; see note 8. The University of Alberta was the host institution for the project, and team membership and structure shifted during the project, with students playing a crucial role in the creation of materials throughout the life of the project; information about team members at the project's completion can be found at http://eugenicsarchive.ca/about#team. Neighborhood Bridges is a human rights advocacy organization based in Edmonton focused on disability culture and directed by Nicola Fairbrother.

10. For the Australian case, see the detailed submission from Women With Disabilities Australia, "Dehumanised: The Forced Sterilisation of Women and Girls with Disabilities in Australia," WWDA Submission to the Senate Inquiry into the Involuntary or Coerced Sterilisation of People with Disabilities in Australia, March 2013, accessed October 1, 2016, http://wwda.org.au/papers/subs/subs2011/. For the Californian prison sterilizations, see Corey Johnson, "Female Inmates Sterilized in California Prisons without Approval," Center for Investigative Reporting, 2013, accessed October 1, 2016, http://cironline.org/reports/female-inmates-sterilized -california-prisons-without-approval-4917. For the earlier California Senate Resolution20, made in 2003, see http://www.csus.edu/cshpe/eugenics/docs/senate _resolution_20.pdf, accessed October 1, 2016. The Chhattisgarh case from India was widely reported in the international media at the end of 2014, including *The Guardian* in the United Kingdom, accessed October 1, 2016, https://www.theguardian .com/world/2014/nov/12/india-sterilisation-deaths-women-forced-camps-relatives; and CNN in the United States, accessed October 1, 2016, http://www.cnn.com/ 2014/11/12/world/asia/india-sterilization-deaths/. As with other cases, it also generated wider reflection on India's history of forced sterilization; see, for example, http://www.bbc.com/news/world-asia-india-30040790, accessed October 1, 2016. For the recent Western Canadian cases, see http://www.cbc.ca/radio/thecurrent/ the-current-for-january-7-2016-1.3393099/aboriginal-women-say-they-were -sterilized-against-their-will-in-hospital-1.3393143, accessed October 1, 2016; and for Peru, see Intercontinental Cry, "Forced Sterilization of 272,000 Indigenous Women 'Not a Crime against Humanity' Public Prosecutor," September 13, 2016, accessed October 1, 2016, https://intercontinentalcry.org/forced-sterilization -272000-indigenous-women-not-crime-humanity-public-prosecutor/.

2   Characterizing Eugenics

1. Hermann Ebbinghaus is best known in psychology for pioneering the experimental study of memory in the late nineteenth century. The "long past, short history" quote begins his popular text, *Psychology: An Elementary Textbook* (Boston: Heath, 1908), published shortly before his death. It became well-known through its repetition in the more widely read history of psychology by Edwin G. Boring, *A History of Experimental Psychology* (New York: Century, 1929). A brief explanation of the neologism "fragile sciences" is given on pages 8–9 of my *Boundaries of the Mind: The Individual in the Fragile Sciences: Cognition* (New York: Cambridge University Press, 2004). For the relevant works of Galton, see note 2 below, and for some standard works on the history of eugenics, see note 4 in chapter 1. Eugenics is sometimes either equated with or assumed to be a form of "social Darwinism" in applying Darwinian ideas to human social problems. I generally avoid that phrase throughout *The Eugenic Mind Project* since in many contexts its connotations seem to me more misleading than helpful.

2. The works of Galton's mentioned here are: "Hereditary Talent and Character," *Macmillan's Magazine* 12 (1865): 157–166, 318–327; *Hereditary Genius: An Inquiry into Its Laws and Consequences* (London: Macmillan, 1869); *English Men of Science: Their Nature and Nurture* (London: Macmillan, 1874); *Inquiries into Human Faculty and Its Development* (London: Macmillan, 1883); and "Race Improvement," in *Memories of My Life* (London: Methuen, 1908), 310–323. Versions of all of these books are available for free download at Internet Archive, accessed October 2, 2017, https:// archive.org/. The quotation given in which Galton coins the term "eugenics" is from a footnote on pages 24–25 of *Inquiries;* the other Galton quotation is from page 323 of *Memories*. For Galton's life and broader contributions to statistics, fingerprinting, and composite portraiture, see Nicholas W. Gillham's *A Life of Sir Francis Galton: From African Exploration to the Birth of Eugenics* (Oxford: Oxford University Press, 2001) and Michael G. Bulmer's *Francis Galton: Pioneer of Heredity and Biometry* (Baltimore, MD: Johns Hopkins University Press, 2003).

3. For a classic discussion of the idea of pure science, see Daniel S. Greenberg, *The Politics of Pure Science* (Chicago: University of Chicago Press, [1967] 1999). For the relationship between statistics and eugenics, see Donald A. MacKenzie, *Statistics in Britain, 1865–1930: The Social Construction of Scientific Knowledge* (Edinburgh: Edinburgh University Press, 1981). Karl Pearson was an accomplished biostatistician who held the first Galton Chair in Eugenics at University College London. In that capacity, Pearson did much to promote eugenics, along with Galton's reputation, including through his editing of *The Life, Letters, and Labours of Francis Galton*, published in three volumes by Cambridge University Press in 1914, 1924, and 1930, a reproduction of which is available at http://galton.org/pearson/, accessed October 2, 2016. The last third of the mathematical biologist Ronald Fisher's *The Genetical Theory of Natural Selection* (Oxford: Clarendon Press, 1930) is dedicated to an

articulation and defense of eugenics. For the idea that science is value-laden and its relationship to the objectivity of science and the ideal of value-freedom, see the broad overview by Julian Reiss and Jan Sprenger, "Scientific Objectivity," *Stanford Encyclopedia of Philosophy* (2014), accessed October 2, 2016, http://plato.stanford. edu/entries/scientific-objectivity/. For representative work on eugenics focused on the United States and discussing immigration and race, see Wendy Kline, *Building a Better Race: Gender, Sexuality, and Eugenics from the Turn of the Century to the Baby Boom* (Berkeley: University of California Press, 2005); Alexandra Minna Stern, *Eugenic Nation: Faults and Frontiers of Better Breeding in Modern America* (Berkeley: University of California Press, 2005); and Edward J. Larson, *Sex, Race, and Science: Eugenics in the Deep South* (Baltimore, MD: Johns Hopkins University Press, 1995). For a detailed, textbook-level discussion of race and eugenics in the United States, see Facing History and Ourselves, *Race and Membership in American History: The Eugenics Movement* (Brookline, MA: Facing History and Ourselves Foundation, 2002); and for a sampling of shorter essays on race and eugenics, see the entries organized under the "Racial Politics" node of Connections and Pathways at EugenicsArchive. ca. For a shorter overview of eugenics and immigration in Canada, see Jacalyn Ambler, "Immigration," EugenicsArchive.ca, and references therein, accessed October 2, 2016, http://eugenicsarchive.ca/discover/connections/53480910132156674b0 002b6. For a broader discussion of race and eugenics, see Marius Turda, "Race, Science, and Eugenics in the Twentieth-Century," in *The Oxford Handbook of the History of Eugenics*, ed. Alison Bashford and Philippa Levine (New York: Oxford University Press, 2010), 62–79, as well as Turda's *Modernism and Eugenics* (Basingstoke: Palgrave, 2010).

4. For eugenics in Germany, including the work of Alfred Ploetz and Friedrich Hertz on racial hygiene in the pre-Nazi era, see Paul Weindling, "German Eugenics and the Wider World: Beyond the Racial State," in *The Oxford Handbook of the History of Eugenics*, ed. Alison Bashford and Philippa Levine (New York: Oxford University Press, 2010), 315–331. For Nazi eugenics, particularly its relationship to medicine, see Robert N. Proctor, *Racial Hygiene: Medicine under the Nazis* (Cambridge, MA: Harvard University Press, 1988); and Paul Weindling, *Health, Race and German Politics between National Unification and Nazism, 1870–1945* (Cambridge: Cambridge University Press, 1989). For Scandinavia, see Gunnar Broberg and Nils Roll-Hansen, eds., *Eugenics and the Welfare State: Sterilization Policy in Denmark, Sweden, Norway, and Finland* (East Lansing: Michigan State University Press, 1996); and the *Scandinavian Journal of History* 24, no. 2 (1999), special issue on "Eugenics in Scandinavia." The extremely useful collection of eugenic family studies assembled by Nicole Rafter is *White Trash: The Eugenic Family Studies 1977–1919* (Boston: Northeastern University Press, 1988); Rafter's introduction to that volume is especially worth reading. My own overview of these studies at EugenicsArchives.ca, "Eugenic Family Studies," relies largely on the material collected in this volume. As intimated previously, much past work on eugenics has concentrated on the United States, and following

and 1931 authored by leading intellectual figures, including the philosopher Bertrand Russell and the biologist J. B. S. Haldane. For an overview of the now largely forgotten series, see Michael Kohlman's "Today and Tomorrow: *To-day and Tomorrow* Book Series," at EugenicsArchive.ca and the links therein to representative samples of books in the series, accessed October 2, 2016, http://eugenicsarchive.ca/discover/connections/546d00a8dabeefbb1a000001.

3　Specifying Eugenic Traits

1. Despite there having been much attention directed at determining *who* were the targets of eugenics and the general categories used to conceptualize those targets (e.g., the unfit, the degenerate), there has been less focus on the kind of questions that structure this chapter. "Eugenic trait" is my own coinage, first introduced in a series of presentations between 2012 and 2015, and discussed more succinctly in my entry "Eugenic Traits" on the EugenicsArchives.ca site. For discussion of "the Unfit," see Elof Axel Carlson, *The Unfit: A History of a Bad Idea* (New York: Cold Spring Harbor Laboratory Press, 2001). On degeneracy, see H. Rimke and A. Hunt, "From Sinners to Degenerates: The Medicalization of Morality in the 19th Century," *History of the Human Sciences* 15, no. 1 (2001): 59–88. For an introduction to the satanic or ritual sexual abuse phenomenon, see Debbie Nathan and Michael Shedeker, *Satan's Silence: Ritual Abuse and the Making of a Modern American Witch Hunt* (New York: Basic Books, 1995); Dorothy Rabinowitz, *No Crueler Tyrannies: Accusation, False Witness, and Other Terrors of Our Times* (Free Press: New York, 2003); and the film *Witch Hunt* (Hard-Nac Movies, Your Half Media Group, Glendale, CA, 2009), accessed June 25, 2017, www.witchhuntmovie.com.

2. For a rich discussion of the Eugenics Record Office on which I have drawn here, see Garland Allen, "The Eugenics Record Office at Cold Spring Harbor, 1910–1940: An Essay in Institutional History," *Osiris* 2 (1986): 225–264; and for the collection of eugenic family studies, see Nicole Rafter, *White Trash: The Eugenic Family Studies, 1877–1919* (Boston: Northeastern University Press, 1988) as well as my "Eugenic Family Studies" (2014), EugenicsArchives.ca. For Galton's original publications, see the notes to chapter 2, and for the early methodological work by Davenport at the ERO, see Charles B. Davenport, Harry H. Laughlin, D. F. Weeks, E. R. Johnstone, and H. H. Goddard, *The Study of Human Heredity: Methods of Collecting, Charting and Analyzing Data*, Eugenics Record Office Bulletin No. 2 (Cold Spring Harbor, NY: Eugenics Record Office, 1911); Charles B. Davenport, *The Trait Book*, Eugenics Record Office Bulletin No. 6 (Cold Spring Harbor, NY: Eugenics Record Office, 1912); and Charles B. Davenport, *The Family History Book*, Eugenics Record Office Bulletin No. 7 (Cold Spring Harbor, NY: Eugenics Record Office, 1912). For a sense of the early influence of the ERO in research and popularization, see the volume by Morton Arnold Aldrich, William Herbert Carruth, Charles Benedict Davenport, Charles Abram Ellwood, Arthur Holmes, William Henry Howell, Harvey Ernest Jordan et al.,

Eugenics: Twelve University Lectures (New York: Dodd, Mead, 1914), especially Davenport's opening lecture, "The Eugenics Programme and Progress in Its Achievement," 1–14. The central organizing role of the notion of social inadequacy within eugenic regimentation is most apparent in Laughlin's model sterilization law; see Harry Hamilton Laughlin, *Eugenical Sterilization in the United States* (Chicago: Psychopathic Laboratory of the Municipal Court of Chicago, 1922), where the Model Law is presented in chapter 15 and discussed in chapter 16. For genealogical representations beyond eugenics, see Mary Bouquet, "Family Trees and Their Affinities: The Visual Imperative of the Genealogical Method," *Man* (n.s.) 2, no. 1 (1996): 43–66; Sandra Bamford and James Leach, eds., *Kinship and Beyond: The Genealogical Model Reconsidered* (New York: Berghann Press, 2009); and J. David Archibald, *Aristotle's Ladder, Darwin's Tree: The Evolution of Visual Metaphors for Biological Order* (New York: Columbia University Press, 2014). For the extension of Morgan's original genealogical work on kinship and its transformation in Australia, leading up to the establishment of its standardization through W. H. R. Rivers, see Patrick McConvell and Helen Gardner, "The Descent of Morgan in Australia: Kinship Representation from the Australian Colonies," *Structure and Dynamics* 6, no. 1 (2013): 1–23, accessed June 25, 2017, http://escholarship.org/uc/item/5711t341; and chapter 10 of their *Southern Anthropology: A History of Fison and Howitt's Kamilaroi and Kurnai* (New York: Palgrave Macmillan, 2015). For Rivers's often-cited standardization, originally published in 1910, see W. H. R. Rivers, "The Genealogical Method of Anthropological Inquiry," in *Kinship and Social Organization* (New York: The Althone Press, [1910] 1968), 97–112. The classic work on *Buck v. Bell* is Paul Lombardo, *Three Generations, No Imbeciles: Eugenics, the Supreme Court, and Buck v. Bell* (Baltimore, MD: Johns Hopkins University Press, 2008). For interesting, recent reflections on the case that focus on its influence and reach, especially in the United States and Germany, see the concluding chapter to Adam Cohen, *Imbeciles: The Supreme Court, American Eugenics, and the Sterilization of Carrie Buck* (New York: Penguin, 2016), 299–323.

3. For an introduction to eugenics and popular culture, see Colette Leung, "Popular Culture," EugenicsArchive.ca, accessed October 10, 2016, http://eugenicsarchive .ca/discover/encyclopedia/535eed7a7095aa000000024a. For discussion of individualistic skew in the fragile sciences, see Robert A. Wilson, *Cartesian Psychology and Physical Minds: Individualism and the Sciences of the Mind* (New York: Cambridge University Press, 1995); *Boundaries of the Mind: The Individual in the Fragile Sciences: Cognition* (New York: Cambridge University Press, 2004), especially chapter 1; and *Genes and the Agents of Life: The Individual in the Fragile Sciences: Biology* (New York: Cambridge University Press, 2005).

4. For marriage restriction laws in the context of American eugenics, see Edward J. Larson, *Sex, Race, and Science: Eugenics in the Deep South* (Baltimore, MD: Johns Hopkins University Press, 1995), especially chapter 2; and Alexandra Minna Stern, *Eugenic Nation: Faults and Frontiers of Better Breeding in Modern America* (Berkeley: University of California Press, 2005), especially chapter 1, and her "Marriage"

(2014), EugenicsArchives.ca, accessed October 10, 2016, http://eugenicsarchive.ca/
discover/encyclopedia/535eeccb7095aa000000023b, from which I have taken the
report of marriage restriction laws between people with disabilities. See also
Michael Billinger, "Miscegenation" (2014), EugenicsArchives.ca, accessed October
10, 2016, http://eugenicsarchive.ca/discover/encyclopedia/52329c0e5c2ec50000000
00b. For the role of Laughlin and the Eugenic Records Office in the Johnson-Reid
Act, see Kenneth M. Ludmerer, *Genetics and American Society: A Historical Approach*
(Baltimore, MD: Johns Hopkins University Press, 1972), chapter 5. For immigration
restriction legislation, particularly in Canada, see Ian Dowbiggin, "'Keeping this
Young Country Sane': C. K. Clarke, Immigration Restriction, and Canadian
Psychiatry, 1890–1925," *The Canadian Historical Review* 76 (1995): 598–627; Jacalyn
Ambler, "Immigration," EugenicsArchive.ca, accessed October 10, 2016, http://
eugenicsarchive.ca/discover/encyclopedia/53480910132156674b0002b6; and Roy
Hanes, "None Is Still Too Many: An Historical Exploration of Canadian Immigration
Legislation as It Pertains to People with Disabilities" (n.d.), Council of Canadians
with Disabilities, http://www.ccdonline.ca/en/socialpolicy/access-inclusion/none
-still-too-many, which contains a discussion of the "undue burden" clause. A copy
of "An Act Respecting Chinese Immigration," passed by the Canadian Parliament on
June 30, 1923, can be found at the Library and Archives Canada website, http://
www.collectionscanada.gc.ca/immigrants/021017-2412.02-e.html. For a recent work
emphasizing the central place of disability and defectiveness in U.S. immigration
history, see Douglas C. Baynton, *Defectives in the Land: Disability and Immigration in
the Age of Eugenics* (Chicago: University of Chicago Press, 2016).

5. The sources that Luke Kersten and I used to compile this data are: Harry Hamilton Laughlin, *Eugenical Sterilization in the United States* (Chicago: Psychopathic
Laboratory of the Municipal Court of Chicago, 1922), accessed October 10, 2016,
https://archive.org/details/cu31924013882109; Lutz Kaelber, *Eugenics: Compulsory
Sterilization in 50 American States* (n.d.), accessed October 10, 2016, http://www.uvm
.edu/%7Elkaelber/eugenics/; J. H. Landman, *Human Sterilization: The History of the
Sexual Sterilization Movement* (New York: Macmillan, 1932); Julius Paul, *"… Three
Generations of Imbeciles Are Enough …" State Eugenic Sterilization Laws in American
Thought and Practice* (1965); Paul Lombardo, *Three Generations, No Imbeciles: Eugenics,
the Supreme Court, and Buck v. Bell* (Baltimore: MD: Johns Hopkins University Press,
2008); and copies of the legislation itself. The digital supplement to Lombardo's
important book contains many significant additional resources here; see www.
buckvbell/othermaterial.html in general and http://buckvbell.com/pdf/JPaulmss.pdf
for the invaluable Julius Paul manuscript.

6. For the text of the original Sexual Sterilization Act of Alberta, see The Alberta
Law Collection, accessed October 10, 2016, http://www.ourfutureourpast.ca/
law/page.aspx?id=2906151; and for brief discussion, see Luke Kersten, "Alberta
Passes Sterilization Act," EugenicsArchive.ca, accessed October 10, 2016, http://
eugenicsarchive.ca/discover/timeline/5172e81ceed5c6000000001d. Likewise, the

1937 amendment to that legislation can be found at http://www.ourfutureourpast .ca/law/page.aspx?id=2968369, accessed October 10, 2016, with discussion by Luke Kersten, "Alberta Passes First Amendment to the Sexual Sterilization Act," EugenicsArchive.ca, accessed October 10, 2016, http://eugenicsarchive.ca/discover/ tree/5172e81ceed5c6000000001d. For a clustering of relevant laws in Alberta, see the "ALBERTA: LAWS" (2014) superordinate node at EugenicsArchives.ca, accessed October 10, 2016, http://eugenicsarchive.ca/discover/connections/535ee4c57095aa 00000001ef. For an important discussion of the legal context of the Alberta sterilization act, see Gerald Robertson's expert witness report, included as Appendix A by Madame Justice Veit in her decision in Muir v. Alberta, 123 Dominion Legal Reports (4th series) (1996), 695–762, 1996 CanLII 7287 (QB AB).

7. Of the resources that I have drawn on in thinking about institutionalization and its relationship to eugenic traits, those of Laughlin and Kaelber listed in note 5 have been most useful; I have drawn particularly on Kaelber for the information reported on Alabama. I have taken the quotation of the 1917 Californian legislation from page 19 of Laughlin's *Eugenical Sterilization in the United States* (Chicago: Psychopathic Laboratory of the Municipal Court of Chicago, 1922). On the broader topic of the institutionalization of people with cognitive disabilities and psychiatric conditions, see Andrew Scull, *Museums of Madness: The Social Organization of Insanity in Nineteenth-Century England* (New York: St. Martin's Press, 1979); James W. Trent, Jr., *Inventing the Feeble Minded: A History of Mental Retardation in the United States* (Berkeley: University of California Press, 1994); and the essays in Roy Porter and David Wright, eds., *The Confinement of the Insane: International Perspectives, 1800– 1965* (Cambridge: Cambridge University Press, 2003). For work that has shaped thinking about institutions and institutionalization, see Erving Goffman, *Asylums: Essays on the Social Situation of Mental Patients and Other Inmates* (New York: Anchor Books, 1961); and Michel Foucault, *History of Madness* (London: Routledge, 2006). And for an accessible introduction to thinking about the relationships between the kinds of institutions exemplified by "training schools" and both Canadian residential schools and schools for the Deaf, see the Institutions module at Eugenics Archives.ca, particularly the submodule "Institutionalization," accessed October 13, 2016, http://eugenicsarchive.ca/discover/institutions.

8. For basic information about the Provincial Training School in Red Deer, Alberta, see Colette Leung, "Provincial Training School" (2014), EugenicsArchives .ca, accessed October 13, 2016, http://eugenicsarchive.ca/discover/institutions/map/ 517da50e9786fa0a73000001. For a brief overview of the under-discussed role of guidance clinics in Alberta's eugenic past, see Amy Samson, "Guidance Clinics" (2014), accessed October 13, 2016, http://eugenicsarchive.ca/discover/encyclopedia/ 555427c835ae9d9e7f000030. For more detailed discussion of guidance clinics, see Amy Samson, "Eugenics in the Community: Gendered Professions and Eugenic Sterilization in Alberta, 1928–1972," *Canadian Bulletin of Medical History* 31, no. 1 (2014): 143–164, and her doctoral dissertation, "Eugenics in the Community:

The United Farm Women of Alberta, Public Health Nursing, Teaching, Social Work, and Sexual Sterilization in Alberta, 1928–1972" (PhD dissertation, University of Saskatchewan, 2014). For the testimony of the eugenics survivors mentioned in the text, see the Our Stories module at the same site, accessed October 13, 2016, http:// eugenicsarchive.ca/discover/our-stories. For recent work on collective memory that provides a useful backdrop for beginning to think about the trauma of a hidden, wrongful past, see Jeffrey K. Olick, Vered Vinitzky-Seroussi, and Daniel Levy, eds., *The Collective Memory Reader* (New York: Oxford University Press, 2011); and for the place of eugenic oral history here, see my "The Role of Oral History in Surviving a Eugenic Past," in *Beyond Testimony and Trauma: Oral History in the Aftermath of Mass Violence*, ed. Steven High (Vancouver: University of British Columbia Press, 2015), 119–138.

9. For the quotation from Philippa Levine and Alison Bashford, see page 6 of their "Introduction: Eugenics and the Modern World," in *The Oxford Handbook of the History of Eugenics* (New York: Oxford University Press, 2010), 3–24. For the work of MacEachran cited, see "Social Legislation in the Province of Alberta," 1934 report to the government of Alberta, University of Alberta Archives, Accession Number 71-217, Item 7, Box 1; the phrase I draw from it is on page 31. For the early work on epilepsy and Huntington's disease associated with Davenport and the Eugenics Records Office, see Charles B. Davenport and David Fairchild Weeks, "A First Study of Inheritance of Epilepsy," *Journal of Nervous and Mental Disease* 38 (1911): 641–670, reprinted as Eugenics Record Office Bulletin No. 4; Smith Ely Jelliffe, Elizabeth B. Muncey, and Charles B. Davenport, "Huntington's Chorea: A Study in Heredity," *Journal of Nervous and Mental Disease* 40, no. 12 (1913): 796–799; Charles B. Davenport, "Huntington's Chorea in Relation to Heredity and Eugenics," *Proceedings of the National Academy of Sciences of the United States of America* 1 (1915): 283–285; Charles B. Davenport and Elizabeth B. Muncey, "Huntington's Chorea in Relation to Heredity and Eugenics," *American Journal of Insanity [Psychiatry]* 73 (1916): 195–222, reprinted as Eugenics Record Office Bulletin No. 17. The report from the American Neurological Association committee referred to is Abraham Myerson, James B. Ayer, Tracy J. Putnam, Clyde E. Keeler, and Leo Alexander, *Eugenical Sterilization—A Reorientation of the Problem. By the Committee of the American Neurological Association for the Investigation of Eugenical Sterilization* (New York: Macmillan, 1936), and the quotations from it are from page 136.

4   Subhumanizing the Targets of Eugenics

1. The full title of the public dialogue I describe was "The Modern Pursuit of Human Perfection: Defining Who Is Worthy of Life," and it was held on October 23, 2008, at the University of Alberta, Edmonton. For the work of Jonathan Glover, see his *What Sort of People Should There Be? Genetic Engineering, Brain Control, and Their Impact on Our Future World* (New York: Penguin Books, 1984) and *Choosing Children:*

Genes, Disability, and Design (Oxford: Clarendon Press, 2006). The article of Nicolas Agar's that I draw on here is "Eugenics, Old and Neoliberal Theories of," in *Encyclopedia of Philosophy and the Social Sciences*, ed. Byron Kaldis, 283–289 (Thousand Oaks, CA: Sage Publications, 2013); the quotation is from page 284. The work I refer to in taking up Agar's point is Alan Buchanan, Dan Brock, Norman Daniels, and Dan Wikler, *From Chance to Choice: Genetics and Justice* (Cambridge: Cambridge University Press, 2000). For a short overview of recent work on what I am calling "subhumanization" see David Livingstone Smith's short entry, "Dehumanization: Psychological Aspects" (2014), EugenicsArchives.ca; for more extended treatment, see his *Less Than Human: Why We Demean, Enslave, and Exterminate Others* (New York: St. Martin's Press, 2011). For overviews emphasizing philosophical and psychological perspectives on dehumanization, respectively, see Smith's "Dehumanization, Essentialism, and Moral Psychology," *Philosophy Compass* 9 (2014): 814–824, and the review provided by Nick Haslam and Stephen Loughnan, "Dehumanization and Infrahumanization," *Annual Review of Psychology* 65 (2014): 399–423.

2. For a powerful testimony of identification with those who lived through Alberta's history of sterilization, see the video narratives of the Alberta artist Nick Supina III at EugenicsArchive.ca. There are two versions of his narrative available, each of which discusses Nick's art and its connection to eugenics, accessed October 23, 2016, http://eugenicsarchive.ca/discover/our-stories/nick and from http://eugenicsarchive .ca/discover/interviews. The second version begins with a discussion of Nick's artwork "Eugenics Painting with Leilani." In support of legal and medical arguments for euthanasia of the disabled, perhaps the most influential work in the history of eugenics that invokes the subhumanizing idea of "life unworthy of life" is Karl Binding and Alfred Hoche's brief *Die Freigabe der Vernichtung Lebensunwerten Lebens: Ihr Mass und Ihr Ziel* (Leipzig: Felix Meiner, 1920). The full text of the second edition is available from Project Gutenberg at http://www.gutenberg.org/ebooks/44565 and from http://chillingeffects.de/binding.pdf, both accessed October 23, 2016; a scholarly translation by Walter E. Wright in English was published as "Permitting the Destruction of Unworthy Life: Its Extent and Form," *Issues in Law and Medicine* 8, no. 2 (1992): 231–265, and an earlier, less scholarly translation with a commentary as *The Release of the Destruction of Life Devoid of Value: It's Measure and It's Form* [sic] (Santa Ana, CA: R. L. Sassone, 1975). The quotations from Troy Duster, *Backdoor to Eugenics* (New York: Routledge, [1990] 2003), are from the preface to the second edition, pages xii and xiv, respectively. For Philip Kitcher's views of the inevitability of eugenics, see his *The Lives to Come: The Genetic Revolution and Human Possibilities* (New York: Simon and Schuster, 1996) and "Utopian Eugenics and Social Inequality," in *Controlling Our Destinies: Historical, Philosophical, Ethical, and Theological Perspectives on the Human Genome Project*, ed. Phillip R. Sloan, (Notre Dame, IN: University of Notre Dame Press, 2000), 229–262, reprinted in Kitcher's *In Mendel's Mirror: Philosophical Reflections on Biology* (New York: Oxford University Press, 2003). For the work of Shakespeare and Agar, see Tom Shakespeare, "The Social Context of

Individual Choice," in *Quality of Life and Human Difference*, ed. David Wasserman, Jerome Bickenbach, and Robert Wachbroit (New York: Cambridge University Press, 2005), 217–236, and *Disability Rights and Wrongs Revisited* (New York: Routledge, 2003); and Nicholas Agar, *Liberal Eugenics: In Defence of Human Enhancement* (Malden, MA: Blackwell, 2004) and *Humanity's End: Why We Should Reject Radical Enhancement* (Cambridge, MA: MIT Press, 2010).

3. For information on Harry J. Haiselden and on the link between eugenics and euthanasia in early twentieth-century North America, see Martin S. Pernick's masterful *The Black Stork: Eugenics and the Death of "Defective" Babies in American Medicine and Motion Pictures Since 1915* (New York: Oxford University Press, 1996), on which I have drawn here; the short quote I give is from page 15. For Martin W. Barr, see his *Mental Defectives: Their History, Treatment and Training* (Philadelphia: P. Blakiston's, 1904) and his "Some Notes on Asexuality; With a Report of Eighteen Cases," *Journal of Nervous and Mental Disease* 51, no. 3 (March 1920): 231–241. The shorter quotation I have used from this article is from page 232; both block quotations that I give are from page 234. Barr's earlier discussion of eugenic sterilization in the book can be found on pages 187–197.

4. For an overview of Le Vann's role in the history of eugenics in Alberta, see Natalie Ball, "Le Vann, Leonard J.," EugenicsArchive.ca, accessed October 23, 2016, http://eugenicsarchive.ca/discover/players/512fa4b134c5399e2c00000d. For the published record of Le Vann's psychotropic experimentation on children at the Provincial Training School of which he was medical superintendent, see Leonard J. Le Vann, "Trifluoperazine Dihydrochloride: An Effective Tranquillizing Agent for Behavioural Abnormalities in Defective Children," *Canadian Medical Association Journal* 80, no. 2 (1959): 123–124; "Thioridazine (Mellaril) A Psycho-sedative Virtually Free of Side-effects," *Alberta Medical Bulletin* 26 (1961): 144–147; "A New Butyrophenone: Trifluperidol: A Psychiatric Evaluation in a Pediatric Setting," *The Canadian Psychiatric Association Journal* 13 (1968): 271–273; "Haloperidol in the Treatment of Behavioral Disorders in Children and Adolescents," *Canadian Psychiatric Association Journal* 14 (1969): 217–220; "Clinical Comparison of Haloperidol and Chlorpromazine in Mentally Retarded Children," *American Journal of Mental* Deficiency 75 (1971): 719–723. For the testimony of Margaret Thompson, see paragraph 114 of the summary of Muir v. Alberta, 123 Dominion Law Reports (4th series) (1996), 695–762, 1996 CanLII 7287 (QB AB), from which the short quotations I give are taken, as well as the account that the lawyer Sandra Anderson gives of her cross-examination of Thompson in my interview with Anderson at EugenicsArchive.ca, accessed October 23, 2016, http://eugenicsarchive.ca/discover/interviews; see especially 19.45–22.15 of the interview, with preceding discussion of Le Vann and following discussion of Thompson. An excerpt from this part of the interview is included in the film *Surviving Eugenics* (Moving Images Distribution, 2015). Thompson's testimony was the chief motivating factor for the 2010 petition, filed by Edmonton

resident Rob Wells, "In the Matter of a Petition for the Removal of a Member of the Order of Canada" (2010) (author's copy) to have Thompson's Order of Canada rescinded by the Governor-General of Canada, a petition that was ultimately denied. For discussion of the petition featuring Rob Wells, see "Margaret Thompson Panel Discussion," accessed October 23, 2016, https://www.youtube.com/watch?v= jpAwWW4wu38, between 8.45–17.45 of the video. There is further documentation of the nature of both Le Vann's medical experimentation on children and on Thompson's role in the history of eugenics in Alberta in sealed records.

5. For the UNICEF Convention on the Rights of the Child (1989), see http:// digitalcommons.ilr.cornell.edu/cgi/viewcontent.cgi?article=1007&context=child, accessed October 23, 2016. On the ratio-centric conception of persons, see Bartlomiej A. Lenart, "Shadow People: Relational Personhood, Extended Diachronic Personal Identity, and Our Moral Obligations toward Fragile Persons" (Ph.D. dissertation, University of Alberta, 2013). For discussion of the alleged subhumanizing tendencies of people with disabilities within moral philosophy, see the essays in Eva Feder Kittay and Licia Carlson, eds., *Cognitive Disability and Its Challenge to Moral Philosophy* (New York: Wiley-Blackwell, 2010). Of special relevance here are the contributions by Peter Singer, "Speciesism and Moral Status," 331–343; Jeff McMahan, "Cognitive Disability and Cognitive Enhancement," 345–367; and Eva Feder Kittay, "The Personal Is Philosophical Is Political: A Philosopher and Mother of a Cognitively Disabled Person Sends Notes from the Battlefield," 393–413. The original talks from the conference on which this volume is based were the subject of a series of blogposts at the Living Archives on Eugenics Blog (at that time, the What Sorts of People Blog), and the thirteen posts there are available from https:// whatsortsofpeople.wordpress.com/2009/02/13/all-wrapped-up-complete-thinking -in-action-series/. Especially notable here are the posts "Peter Singer on Parental Choice, Disability, and Ashley X," "Singer's Assault on Universal Human Rights," and "Peter Singer and Profound Intellectual Disability" by the disability studies scholar Dick Sobsey, and the surrounding discussion and commentaries. For a reflection on persons, political theory, and intellectual disability that utilizes Judith Butler's notion of normative violence, see Ashley Taylor, "Expressions of 'Lives Worth Living' and Their Foreclosure through Philosophical Theorizing on Moral Status and Intellectual Disability," in *Foucault and the Government of Disability*, rev. ed., ed. Shelley Tremain (Ann Arbor: University of Michigan Press, 2015), 372–395. For the quotation from Harriet McBryde-Johnson and her account of her interactions with Singer, see her "Unspeakable Conversations," *New York Times Magazine*, February 16, 2003. For the views of Julian Savulescu, see his "Procreative Beneficence: Why We Should Select the Best Children," *Bioethics* 15 (5/6) (2001): 413–426, and "Procreative Beneficence: Reasons to Not Have Disabled Children," in *The Sorting Society: The Ethics of Genetic Screening and Therapy* ed. Loane Skene and Janna Thompson (New York: Cambridge University Press, 2008), 51–68; and Julian Savulescu and Guy Kahane, "The Moral Obligation to Create Children with the Best Chance of the Best

Life," *Bioethics* 23, no. 5 (2009): 274–290. The cited principle of procreative beneficence is stated at page 415 of the earliest of these papers; it is re-expressed in the joint paper from 2009. Savulescu discusses opportunity restriction on pages 63–66 of the 2008 paper and extends the reach of procreative beneficence in both later papers; the final quotation from Savulescu in this section is from page 66.

6. For a pair of interesting discussions of Ashley X that includes references to much of the literature, see Alison Kafer, "At the Same Time, Out of Time: Ashley X," chapter 2 of her *Feminist, Queer, Crip* (Bloomington: Indiana University Press, 2013), 47–68; and William J. Peace and Claire Roy, "Scrutinizing Ashley X: Presumed Medical 'Solutions' vs. Real Social Adaptation," *Journal of Philosophy, Science, and Law* 14, no. 3 (July 2014): 33–52. The original paper reporting the case medically is Daniel F. Gunther and Douglas S. Diekema, "Attenuating Growth in Children with Profound Developmental Disability: A New Approach to an Old Dilemma," *Archives of Pediatrics and Adolescent Medicine* 160, no. 10 (2006): 1013–1017; the quotation I give from that paper is on page 1013. For an early report that was critical of the treatment, see "Investigative Report Regarding the 'Ashley Treatment,'" Disability Rights Education and Defense Fund, May 2007, accessed October 23, 2016, https:// dredf.org/public-policy/ethics/investigative-report-regarding-the-ashley-treatment/. The report in *The Guardian* on the expansion in the treatment is from March 16, 2012, accessed October 23, 2016, http://www.guardian.co.uk/society/2012/mar/16/ growth-attenuation-treatment-toms-story. For introductory discussion of the veil of ignorance as a distinguishing feature of John Rawls's theory of justice within the social contract tradition, see Samuel Freeman, "Original Position," *Stanford Encyclopedia of Philosophy* (2014), especially sections 1–3, accessed October 23, 2016, http:// plato.stanford.edu/entries/original-position/. For discussion of disability and theories of justice, including Rawls's, see David Wasserman, Adrienne Asch, Jeffrey Blustein, and Daniel Putnam, "Disability and Justice," *Stanford Encyclopedia of Philosophy* (2013), accessed October 23, 2016, http://plato.stanford.edu/entries/disability-justice/. The quotation from Eva Feder Kittay is part of her dissenting opinion in a group report on the Ashley case: Benjamin S. Wilfond, Paul Steven Miller, Carolyn Korfiatis, Douglas S. Diekema, Denise M. Dudzinski, and Sara Goering, "Navigating Growth Attenuation in Children with Profound Disabilities," *Hastings Center Report* 40, no. 6 (2010): 27–40, at 32.

7. On Project Value, see the Canadian Association for Community Living, "Project Value: Disabled Lives Have Value" (2016), accessed October 23, 2016, http://www .cacl.ca/news-stories/blog/project-value-disabled-lives-have-value. The video narratives themselves, of which more than twenty were posted in a three-month period, continue to be developed, having started in late July 2016. The text of Bill C-14 can be read in French and English at Government of Canada (2016), http://www.parl .gc.ca/HousePublications/Publication.aspx?Language=E&Mode=1&DocId=8384014, with a summary characterization of it at https://openparliament.ca/bills/42-1/C-14/, both accessed October 23, 2016.

5 Where Do Ideas of Human Variation Come From?

1. For a recent collection of work on prosociality, see David A. Schroeder and William G. Grazziano, eds., *The Oxford Handbook of Prosociality* (New York: Oxford University Press, 2015).

2. Discussions of human variation have been dominated by a focus on race and ethnicity, where the related question of the origins or racial and ethnic categories has centered on the objectivity of those categories and their relationship to science. For example, see the population geneticist Richard Lewontin's *Human Diversity* (New York: Scientific American Library, 1982) and the anthropologist Stephen Molnar's *Human Variation: Races, Types, and Ethnic Groups* (New York: Routledge, 2006, 6th ed.). See also Michael Billinger's "Ethnicity and Race," EugenicsArchive.ca, accessed November 10, 2016, http://eugenicsarchive.ca/discover/connections/5508686c5f8ec c9f71000003.

3. For interesting work on the epistemology of disablement, reflecting the contrasting, lived experience of disablement, see Susan Wendell, *The Rejected Body: Feminist Reflections on Disability* (New York: Routledge, 1996); and Eli Clare, *Exile and Pride: Disability, Queerness, and Liberation* (Durham, NC: Duke University Press, 2009). Although it has not been expressed in terms of the notion of marked variation and has taken up disability only marginally, the recent literature on implicit attitudes and implicit bias, anchored in a longer tradition of social psychology focused on in-group and out-group identification, is informative in thinking about the phenomenology of marked variation and disablement. For an introduction and overview, see Michael Brownstein, "Implicit Bias," *Stanford Encyclopedia of Philosophy* (Spring 2016 edition), ed. Edward N. Zalta, accessed November 10, 2016, https://plato.stanford .edu/archives/spr2016/entries/implicit-bias/; and for more detailed discussions, see the essays in Michael Brownstein and Jennifer Saul, eds., *Implicit Bias and Philosophy: Volume 2, Moral Responsibility, Structural Injustice, and Ethics* (Oxford: Oxford University Press, 2016).

4. For the ICIDH definitions of impairment, disability, and handicap, see World Health Organization, *International Classification of Impairments, Disabilities, and Handicaps* (Geneva: World Health Organization, [1980] 1993), accessed November 10, 2016, http://apps.who.int/iris/bitstream/10665/41003/1/9241541261_eng.pdf. The definitions quoted can be found on pages 27–29, and the further characterization of handicap that I cite is also from page 29. The concept of normalcy has a long-standing history and is pervasive in discussions of disease, medical treatment, and disability. See the classic discussion of Georges Canguilhem, *On the Normal and the Pathological* (Dordrecht: D. Reidel [1966] 1978) and the more recent critique of Anita Silvers, "A Fatal Attraction to Normalizing," in *Enhancing Human Traits: Ethical and Social Implications*, ed. Erik Parens (Washington, DC: Georgetown University Press, 1998), 95–121. For more recent discussion of disability and normalizing,

particularly in the medical and health sciences, see Anita Silvers, "Disability and Normality," in *Routledge Companion to Philosophy of Medicine*, ed. Miriam Solomon, Jeremy R. Simon, and Harold Kincaid (New York: Routledge, 2016), 36–47.

5. Note that all in-text page references in this section are to Lennard Davis's *Enforcing Normalcy*. Foucault's work on normalcy is diverse and wide-ranging, but see, for example, his *Madness and Civilization* (New York: Random House, 1965; translated by Richard Howard), *Discipline and Punish: The Birth of the Prison* (New York: Vintage, 1979; translated by Alan Sheridan); *Abnormal: Lectures at the College de France 1974–75* (New York: Picador, 2003; translated by Graham Burchell), and *"Society Must Be Defended": Lectures at the College de France 1975–76* (New York: Picador, 2003; translated by David Macey). The works of Davis, Tremain, and Rose that I refer to are Lennard Davis, *Enforcing Normalcy: Deafness, Disability and the Body* (London: Verso, 1995) and his collection *The Disability Studies Reader*, 5th edition (New York: Routledge, [1995] 2016); Shelley Tremain, "On the Government of Disability," *Social Theory and Practice* 27, no. 4 (2001): 617–636; and Nikolas Rose, *The Psychological Complex: Psychology, Politics and Society in England 1969–1939* (London: Routledge and Kegan Paul, 1985). For the Foucauldian perspective on disability more generally, see the essays in Shelley Tremain, ed., *Foucault and the Government of Disability*, rev. ed. (Ann Arbor: University of Michigan Press, [2005] 2015). The work of MacKenzie on which Davis draws is Donald A. MacKenzie, *Statistics in Britain: The Social Construction of Scientific Knowledge, 1865–1930* (Edinburgh: Edinburgh University Press, 1981).

6. For the more recent work of Nikolas Rose, see his *The Politics of Life Itself: Biomedicine, Power, and Subjectivity in the Twenty-First Century* (Princeton, NJ: Princeton University Press, 2007) and Nikolas Rose and Joelle M. Abi-Rached, *Neuro: The New Brain Sciences and the Management of the Mind* (Princeton, NJ: Princeton University Press, 2012). For the quote from Aristotle, see book VII, 16, of *The Politics* (1335b 20–21) in *The Complete Works of Aristotle: The Revised Oxford Translation*, vol. 2, ed. Jonathan Barnes (Princeton, NJ: Princeton University Press, 1984). The reference to G. E. R. Lloyd is to his *Being, Humanity, and Understanding* (New York: Oxford University Press, 2012). My brief remarks on deafness and disability in the ancient world have relied on Martha L. Edwards, "Deaf and Dumb in Ancient Greece," in *The Disability Studies Reader*, 2nd edition, ed. Lennard Davis (New York: Routledge, 1997), 29–51, and "Constructions of Physical Disability in the Ancient World—The Community Concept," in *The Body and Physical Difference: Discourses of Disability*, ed. David T. Mitchell and Sharon L. Snyder (Ann Arbor: University of Michigan Press, 1997), 35–50.

7. For G. E. Moore's original discussion of the open question argument, see his *Principia Ethica* (New York: Cambridge University Press, 1903); and for a succinct overview of some of the complexities of, and significance to, the open question argument in the history of metaethics, see Michael Ridge, "Moral Non-naturalism," *Stanford*

Encyclopedia of Philosophy (2014), especially sections 1–2, accessed November 11, 2016, https://plato.stanford.edu/entries/moral-non-naturalism/.

6 A Socio-cognitive Framework for Marked Variation

1. For Hobbes's own views, see his *Leviathan* (Cambridge: Cambridge University Press, [1651] 1991). The reference to C. B. Macpherson is to his *The Political Theory of Possessive Individualism* (Oxford: Clarendon Press, 1962). For a brief overview of Hobbes's political philosophy and social contract theory, see Sharon A. Lloyd and Susanne Sreedhar, "Hobbes's Moral and Political Philosophy," *Stanford Encyclopedia of Philosophy* (Spring 2014 edition), ed. Edward N. Zalta, accessed November 20, 2016, https://plato.stanford.edu/archives/spr2014/entries/hobbes-moral/; and for expressions of the standard view, see Richard Tuck, *Hobbes* (Oxford: Oxford University Press, 1989); J. W. N. Watkins, *Hobbes's System of Ideas* (London: Hutchison, 1965); and Peter Zagorin, *Hobbes and the Law of Nature* (Princeton, NJ: Princeton University Press, 2009). My own views of Hobbes have been influenced by the work of R. E. Ewin; see his *Cooperation and Human Values: A Study in Moral Reasoning* (New York: St. Martin's Press, 1981) and *Virtues and Rights: The Moral Philosophy of Thomas Hobbes* (Boulder, CO: Westview Press, 1991).

2. The most wide-ranging and influential work on nonhuman sociality in the past fifty years is Edward O. Wilson's *Sociobiology: The New Synthesis* (Cambridge, MA: Belknap Press, 1975). For some speculative and provocative recent thoughts about sensory systems and the mobile living world, see Malcolm MacIver, "Neuroethology: From Morphological Computation to Planning," in *The Cambridge Handbook of Situated Cognition*, ed. Philip Robbins and Murat Aydede (New York: Cambridge University Press, 2009), 480–504, and what MacIver calls the "Buena Vista Sensing Club Hypothesis"; for MacIver's most recent development of this idea, see Malcolm A. MacIver, Lars Schmitz, Ugurcan Mugan, Todd D. Murphey, and Curtis D. Mobley, "Massive Increase in Visual Range Preceded the Origin of Terrestrial Vertebrates," *Proceedings of the National Academy of Sciences* 114, no. 12 (2017): E2375–E2384.

3. For the Machiavellian intelligence hypothesis, see Richard Byrne and Andrew Whiten, eds., *Machiavellian Intelligence: Social Expertise and the Evolution of Intellect in Monkeys, Apes, and Humans* (New York: Oxford University Press, 1989), and their *Machiavellian Intelligence II: Extensions and Evaluations* (New York: Oxford University Press, 1997). For departures from the general, individualistic view of cognitive sophistication that has been predominant in the cognitive sciences, see my *Boundaries of the Mind* (New York: Cambridge University Press, 2004); Robert A. Wilson and Andy Clark, "How to Situate Cognition: Letting Nature Take Its Course," in *The Cambridge Handbook of Situated Cognition*, ed. Philip Robbins and Murat Aydede (New York: Cambridge University Press, 2009), 55–77; and the essays in Richard Menary, ed., *The Extended Mind* (Cambridge, MA: MIT Press, 2010). For

the relationship between embodied and extended cognition, see Andy Clark, *Supersizing the Mind: Embodiment, Action, and Cognitive Extension* (New York: Oxford University Press, 2008); and Robert A. Wilson and Lucia Foglia, "Embodied Cognition," *Stanford Encyclopedia of Philosophy* (Winter 2016 edition), ed. Edward N. Zalta, https://plato.stanford.edu/archives/win2016/entries/embodied-cognition/. For work in this tradition on visual perception, see Alva Noë, *Action in Perception* (Cambridge, MA: MIT Press, 2004); and Robert A. Wilson, "Extended Vision," in *Perception, Action and Consciousness: Sensorimotor Dynamics and Two Visual Systems*, ed. Nivedita Gangopadhyay, Michael Madary, and Finn Spicer (New York: Oxford University Press, 2010), 277–290; on moral cognition, see Andrew Sneddon, *Like-Minded: Externalism and Moral Psychology* (Cambridge, MA: MIT Press, 2011); on emotions and emotional development, see Jennifer Greenwood, *Becoming Human: The Ontogenesis, Metaphysics, and Expression of Human Emotionality* (Cambridge, MA: MIT Press, 2015); on musical cognition, see Joel Krueger, "Affordances and the Musically Extended Mind," *Frontiers of Psychology*, 4, no. 1003 (2014): 1–9; and Luke Kersten and Robert A. Wilson, "The Sound of Music, Externalist Style," *American Philosophical Quarterly* 53, no. 2 (2016): 139–154. For the expression of externalism in terms of the notion of cognitive resources, see my "Ten Questions Concerning Extended Cognition," *Philosophical Psychology* 27, no. 1 (2014): 19–33. For important criticisms of the idea of the extended mind and extended cognition, see Frederick Adams and Ken Aizawa, *The Bounds of Cognition* (New York: Blackwell, 2008); and Robert Rupert, *Cognitive Systems and the Extended Mind* (New York: Oxford University Press, 2009). For Durkheim's classic appeal to collective representations, see Emile Durkheim, "Individual and Collective Representations," in *Sociology and Philosophy* (New York: Simon and Schuster, [1924] 1974; originally published in French), 1–34; the essay itself was originally published in French in 1898; for useful discussion see Marcel Fournier, "Durkheim's Life and Context: Something New About Durkheim?," in *The Cambridge Companion to Durkheim*, ed. Jeffrey C. Alexander and Philip Smith (New York: Cambridge University Press, 2005), 41–69, esp. 55–60. I discuss the broader collective psychology tradition in chapters 11 and 12 of *Boundaries of the Mind*.

4. For influential work on collective intentionality, see John R. Searle, *The Construction of Social Reality* (New York: Free Press, 1995) and *Making the Social World: The Structure of Human Civilization* (New York: Oxford University Press, 2010); Margaret Gilbert, *Living Together: Rationality, Sociality, and Obligation* (New York: Rowman and Littlefield, 1996); Michael Tomasello, *A Natural History of Human Thinking* (Cambridge, MA: Harvard University Press, 2014); and Raimo Tuomela, *The Philosophy of Sociality: The Shared Point of View* (New York: Oxford University Press, 2007). For a development of my views of collective intentionality, see my "Collective Intentionality in Non-human Animals," in *Routledge Handbook of Collective Intentionality*, ed. Marija Jankovic and Kirk Ludwig (New York: Routledge, 2017), and "Group-level Cognizing, Collaborative Remembering, and Individuals," in *Collaborative Remem-*

bering: How Remembering with Others Influences Memory, ed. Michelle Meade, Amanda Barnier, Penny Van Bergen, Celia Harris, and John Sutton (New York: Oxford University Press, 2017).

5. For my recent thinking on nonhuman animal cognition, see the works of mine referred to in note 4; for my earlier thoughts about animal cognition and collective intentionality, see my "Social Reality and Institutional Facts: Sociality Within and Without Intentionality," in *Intentional Acts and Institutional Facts: Essays on John Searle's Social Ontology*, ed. Savas L. Tsohatzidis (Dordrecht: Springer, 2007), 139–153. For a recent, accessible view of nonhuman cognition, see Peter Godfrey-Smith, *Other Minds: The Octopus, the Sea, and the Deep Origins of Consciousness* (New York: Farrar, Strauss, and Giroux, 2016).

6. For Gil-White's views of the evolutionary psychology of ethnicity, see his "Are Ethnic Groups Biological 'Species' to the Human Brain?," *Current Anthropology* 42, no. 4 (2001): 515–553; and for a developmental psychology approach that appeals to something like a living kinds module, see Frank C. Keil, *Concepts, Kinds, and Cognitive Development* (Cambridge, MA: MIT Press, 1989). For the distinction between proper and actual domains for modules, see Dan Sperber, *Explaining Culture: A Naturalistic Approach* (New York: Cambridge University Press, 1996). For the original infant imitation study, see Andrew N. Meltzoff and N. Keith Moore, "Imitation of Facial and Manual Gestures by Human Neonates," *Science* 198, no. 4312 (1977): 75–78; for Meltzoff's more explicit thoughts about "like me" detectors see his "Like Me: A Foundation for Social Cognition," *Developmental Science* 19, no. 1 (2007): 126–134; and for a recent challenge to Meltzoff's imitation paradigm, see Janine Oostenbroek, Thomas Suddendorf, Mark Nielsen, Jonathan Redshaw, Siobhan Kennedy-Costantini, Jacqueline Davis, Sally Clark, and Virginia Slaughter, "Comprehensive Longitudinal Study Challenges the Existence of Neonatal Imitation in Humans," *Current Biology* 26, no. 10 (2016): 1334–1338.

7. Much of my knowledge about thalidomide and thalidomiders comes from many stimulating discussions with Gregor Wolbring. For the view of parenting thalidomiders conveyed here, see Ethel Roskies, *Abnormality and Normality: The Mothering of Thalidomide Children* (Ithaca, NY: Cornell University Press, 1972); and T. Stephens and R. Brynner, *The Dark Remedy: The Impact of Thalidomide* (New York: Perseus Publishing, 2001). For Gregor Wolbring's brief reflections on his own upbringing, see his "Parents Without Prejudice," in *Reflections from a Different Journey: What Adults with Disabilities Wish All Parents Knew*, ed. Stanley Klein and John Kemp (New York: McGraw Hill, 2004), 18–22; and for more extended personal reflections on eugenics, genetic testing, and thalidomide, see the first ten minutes or so of his interview at Interviews+ at EugenicsArchive, accessed November 20, 2016, http://eugenicsarchive.ca/discover/interviews. I discuss the application to kinship of what I am here calling the "prosocial flipside" in more detail in a book nearing completion, *Relative Beings* (unpublished manuscript).

7   Back Doors, Newgenics, and Eugenics Underground

1. For the quotation from Nicholas Agar, see his *Liberal Eugenics: In Defence of Human Enhancement* (Malden, MA: Blackwell, 2004), 7.

2. For a discussion of the long-term outcomes of prenatal testing for trisomy 21, see Brian G. Skotko, "With New Prenatal Testing, Will Babies with Down Syndrome Slowly Disappear?," *Archives of Disease in Childhood* 94 (2009): 823–826. For the quotation from Marsha Saxton, see her "Disability Rights and Selective Abortion," in *Abortion Wars: A Half Century of Struggle, 1950–2000* ed. Rickie Solinger (Berkeley: University of California Press, 1997), 374–395, at 391. For Down syndrome, including work on quality of life and well-being, see R. Brown, J. Taylor, and B. Matthews, "Quality of Life—Ageing and Down Syndrome," *Down Syndrome Research and Practice* 6, no. 3 (2001): 111–116; Jan Gothard, *Greater Expectations: Living with Down Syndrome in the 21st Century* (Fremantle: Fremantle Press, 2011); and Michael Berube, "Down Syndrome," April 28, 2014, accessed November 23, 2016, http://www .eugenicsarchive.ca/discover/encyclopedia/535eeb507095aa000000021d. For termination rates, respectively, in Europe and North America, see P. Boyd, C. DeVigan, B. Khoshnood, M. Loane, E. Garne, H. Dolk, and the EUROCAT working group, "Survey of Prenatal Screening Policies in Europe for Structural Malformations and Chromosome Anomalies, and Their Impact on Detection and Termination Rates for Neural Tube Defects and Down's Syndrome," *BJOG: An International Journal of Obstetrics and Gynaecology* 115 (2008): 689–696; and Jamie L. Natoli, Deborah L Ackerman, Suzanne McDermott, and Janice G. Edwards, "Prenatal Diagnosis of Down Syndrome: A Systematic Review of Termination Rates (1995–2011)," *Prenatal Diagnosis* 32, no. 2 (2013): 142–153. For the Society of Obstetricians and Gynaecologists of Canada 2007 guidelines, see "Prenatal Screening for Fetal Aneuploidy," *Journal of Obstetricians and Gynaecologists of Canada* 187 (February 2007): 146–161; the first quote I have given is from page 146 and the second from the abstract. See also the more recent committee opinion, "Counselling Considerations for Prenatal Genetic Screening," *Journal of Obstetricians and Gynaecologists of Canada* 277 (May 2012): 489–493. An early characterization of prenatal screening and testing as "search and destroy missions" associated with Koop can be found in Bernard Weinraub, "Reagan Nominee for Surgeon General Runs into Obstacles on Capitol Hill," *New York Times*, April 7, 1981, A16/6. Koop's general, religion-based anti-abortion views motivate his characterization; for example, see C. Everett Koop and Francis A. Schaeffer, *Whatever Happened to the Human Race* (Westchester, IL: Crossway Books, 1979). The phrase "search and destroy missions" was revived sixteen years later in the popular media by George Will in an influential *Newsweek* article in 2007; see George Will, "Golly, What *Did* Jon Do?," *Newsweek*, January 29, 2007. Prenatal genetic diagnosis was developed by Santiago Munné in his doctoral work at Cornell in the late 1980s; he now heads the company Reprogenetics, http://reprogenetics.com/, accessed November 23, 2016. The bioethicist Rob Sparrow coins and discusses "in vitro eugenics" in an article of that name in *Journal of Medical Ethics* 40, no. 11 (2014): 725–731.

3. For the quotation from Frederick Osborn, see his *Preface to Eugenics* (New York: Harper and Brothers, 1940), 296–297; the shorter quote from James Watson is from his *A Passion for DNA: Genes, Genomes, and Society* (Cold Spring Harbor, NY: Cold Spring Harbor Laboratory Press, 2001), 228. The idea of eugenic logic is introduced in Rosemarie Garland-Thomson, "The Case for Conserving Disability," *Journal of Bioethical Inquiry* 9, no. 3 (2012): 339–355; the quotations are from 339–340. For disability as the master trope of human disqualification, see David Mitchell and Laura Snyder, "The Eugenic Atlantic: Race, Disability, and the Making of an International Eugenic Science, 1800–1945," *Disability and Society* 18, no. 7 (2003): 843–864, at 861.

4. For an overview of expressivism and the disability rights critique, see Erik Parens and Adrienne Asch, "Disability Rights Critique of Prenatal Genetic Testing: Reflections and Recommendations," in *Prenatal Testing and Disability Rights,* ed. Erik Parens and Adrienne Asch (Washington, DC: Georgetown University Press, 2000), 3–43, originally published in *Hastings Center Report* (Sept.–Oct. 1999), S1–22. For discussion of pregnancy as a conflict of rights and a brief introduction to fetal rights beyond the right to life, see Ruth Hubbard, "Personal Courage Is Not Enough: Some Hazards of Childbearing in the 1980s," in *Test-Tube Women: What Future for Motherhood,* ed. Rita Arditti, Renate Duelli Klein, and Shelley Minden (London: Pandora Press, 1984), 331–355. The Saxton quotes come, respectively, from her "Disability Rights and Selective Abortion," in *Abortion Wars: A Half Century of Struggle, 1950–2000,* ed. Rickie Solinger (Berkeley: University of California Press, 1997), 374–395, at 391, and "Why Members of the Disability Community Oppose Prenatal Diagnosis and Selective Abortion," in *Prenatal Testing and Disability Rights,* ed. Erik Parens and Adrienne Asch, cited in this note, 147–164, at 147; see also her "Born and Unborn: The Implications of Reproductive Technologies for People with Disabilities," also in *Test-Tube Women*, 298–312. Asch's influential and broad work on expressivism and the disability rights critique of prenatal screening is rooted in her more general thinking about reproductive technology; for example, see her "Reproductive Technology and Disability," in *Reproductive Laws for the 1990s,* ed. Sherrill Cohen and Nadine Taub (Clifton, NJ: Humana Press, 1989), 69–124, and Adrienne Asch and Michelle Fine, eds., *Women with Disabilities: Essays in Psychology, Culture, and Politics* (Philadelphia, PA: Temple University Press, 1988). For two of the more influential, direct contributions to the disability rights critique that Asch has made, see her "Why I Haven't Changed My Mind about Prenatal Diagnosis: Reflections and Refinements," reprinted in *Prenatal Testing and Disability Rights,* ed. Erik Parens and Adrienne Asch, cited above, as well as her "Disability Equality and Prenatal Testing: Contradictory or Compatible?," *Florida State University Law Review* 30 (2003): 315–342.

5. The quotations from Savulescu on opportunity restriction and disability are from his 2008 paper listed below. See the works cited in notes 1 and 2 for Agar's liberal eugenics and Sparrow's in vitro eugenics; the quotations from Agar are drawn from page 4 and 5 of his *Liberal Eugenics*. For utopian eugenics, see Philip Kitcher, "Utopian Eugenics and Social Inequality," in *In Mendel's Mirror: Philosophical Reflections*

on Biology (New York: Oxford University Press, 2003), 258–282, originally published in *Controlling Our Destinies: Historical, Philosophical, Ethical, and Theological Perspectives on the Human Genome Project*, ed. Philip R. Sloan (Notre Dame, IN: University of Notre Dame Press, 2000), 229–262. For Agar's more recent thought, see his "Eugenics, Old and Neoliberal Theories of," in *Encyclopedia of Philosophy and the Social Sciences*, ed. Byron Kaldis (Thousand Oaks, CA: Sage Publications), 283–289; the longer quotation is from 287. For the original statement and defense of the Principle of Procreative Beneficence, see Julian Savulescu, "Procreative Beneficence: Why We Should Select the Best Children," *Bioethics* 15, no. 5/6 (2001): 413–426, and for a follow-up, see Julian Savulescu, "Procreative Beneficence: Reasons Not to Have Disabled Children," in *The Sorting Society: The Ethics of Genetic Screening and Therapy*, ed. J. Thompson and L. Skene (Cambridge: Cambridge University Press, 2008), 51–68. The coauthored paper from which all quotations in my discussion of Procreative Beneficence here are taken is Julian Savulescu and Guy Kahane, "The Moral Obligation to Create Children with the Best Chance of the Best Life," *Bioethics* 23, no. 5 (2009): 274–290. For critiques of the principle on the ground cited in the text, see Adrienne Asch and David Wasserman, "Where Is the Sin in Synecdoche?: Prenatal Testing and the Parent-Child Relationship," in *Quality of Life and Human Difference: Genetic Testing, Health Care, and Disability*, ed. David Wasserman, Jerome Bickenbach, and Robert Wachbroit (New York: Cambridge University Press, 2005), 172–216; P. Herissone-Kelly, "Procreative Beneficence and the Prospective Parent," *Journal of Medical Ethics* 32, no. 3 (2006): 166–169; and Wasserman's contribution to David Benatar and David Wasserman, *Debating Procreation: Is It Wrong to Reproduce?* (New York: Oxford University Press, 2015). The dilemma argument for those aiming to use Procreative Beneficence to derive conclusions about disability and well-being, as well as discussion of Disability-Free Procreation, that is summarized in this section is articulated in full in Matthew J. Barker and Robert A. Wilson, "Well-Being, Disability, and Choosing Children" (unpublished manuscript).

6. Lennard Davis's recent views about normalcy can be found in his *The End of Normalcy: Identity in a Biocultural Era* (Lansing: University of Michigan Press, 2013); the quotations given from that book in the second last paragraph in this section are from pages 9 and 4, respectively, and his discussion of the hyper-marginalized is on page 13. For an example of the continuing appeal to normalcy in discussions of eugenics, perhaps reflecting Davis's own earlier influence, see Bill Armer, "Eugenetics: A Polemical View of Social Policy in the Genetic Age," *New Formations* 60 (2007): 89–101. On the medical-industrial complex, see Barbara Ehrenreich and John Ehrenreich, *The American Health Empire: Power, Profits, and Politics* (New York: Vintage, 1971); and for the more recent argument about it that I have appealed to, see Carroll L. Estes, Charlene Harrington, and David N. Pellow, "Medical-Industrial Complex," in *Encyclopedia of Sociology*, 2nd edition, vol. 3, ed. Edgar F. Borgatta and Rhonda J. V. Montgomery (New York: Macmillan, 2000), 1818–1832.

7. For "debility" and neoliberalism, see Jasbir Puar, "The Cost of Getting Better: Ability and Debility," in *The Disability Studies Reader*, 4th edition, ed. Lennard Davis (New York: Routledge, 2013), 177–184; the quotations are from pages 182 and 180, respectively. For Haraway's discussion of the shift from perfection to optimization, see Donna Haraway, "A Cyborg Manifesto: Science, Technology, and Socialist-Feminism in the Late Twentieth Century," in her *Simians, Cyborgs and Women: The Reinvention of Nature* (New York: Routledge, 1991), 149–181. On the interplay between notions of health and disability and social oppression, see James I. Charleton, *Nothing about Us without Us: Disability Oppression and Empowerment* (Berkeley: University of California Press, 1998); and for more recent reflections on disability and institutionalization, see Liat Ben-Moshe, Chris Chapman, and Allison C. Carey, eds., *Disability Incarcerated: Imprisonment and Disability in the United States and Canada* (New York: Palgrave Macmillan, 2014). For disability and the stratification of socio-economic inequality through individual cost, see the work by Puar already cited, as well as David U. Himmelstein, Deborah Thorne, Elizabeth Warren, and Steffie Woolhandler, "Medical Bankruptcy in the United States, 2007: Results of a National Study," *American Journal of Medicine* 122 (2009): 741–746. The quotation from Savulescu is from his "Procreative Beneficence: Why We Should Select the Best Children," *Bioethics* 15, no. 5/6 (2001): 413–426, at 424; that from Savulescu and Kahane is from their "The Moral Obligation to Create Children with the Best Chance of the Best Life," *Bioethics* 23, no. 5 (2009): 274–290, at 290. For Judith Butler's concept of responsibilization, see her *Frames of War: When Is Life Grievable?* (New York: Verso, 2009). Angela Davis's comment was made during a public lecture at the University of Alberta, Edmonton, Canada, "On Prisons, Race, and Gender Based Violence," March 15, 2014.

8. A sense of the dialectic between Autism Speaks and autistic self-advocates can perhaps be best gained through online blogging exchanges. For a representative view of Autism Speaks, see "What Is Asperger Syndrome/HFA?" (2010), https://www.autismspeaks.org/family-services/tool-kits/asperger-syndrome-and-high-functioning-autism-tool-kit/introduction; for the quotations from its president, Suzanne Wright, see "Autism Speaks to Washington—A Call for Action" (2013), https://www.autismspeaks.org/news/news-item/autism-speaks-washington-call-action. For an early response to Autism Speaks from the Autistic Self-Advocacy Network, see Meg Evans, "Why Autism Speaks Does Not Speak For Us" (2009), Autistic Self-Advocacy Network, Central Ohio, http://asancentralohio.blogspot.ca/2009/08/why-autism-speaks-does-not-speak-for-us.html; more recently, see Autistic Advocacy, "Before You Donate to Autism Speaks, Consider the Facts" (2016), https://autisticadvocacy.org/wp-content/uploads/2016/03/AutismSpeaksFlyer_color_2016.pdf. For responses to Autism Speaks from individual autistic bloggers, see The Caffeinated Autistic, "New Autism Speaks Masterpost" (2014), https://thecaffeinated autistic.wordpress.com/new-autism-speaks-masterpost-updated-62014/; Alyssa Hillary, "Autism-Speaks Are Work-Stealing, White-Texting Liars," *Yes, That Too* (2014), http://yesthattoo.blogspot.ca/2014/01/autism-speaks-are-work-stealing-white.html;

and Autistic Chick, "Nonspeaking. Real. Self-Advocates" (2014), http://autisticchick
.blogspot.ca/2014/08/i-left-comment-on-thing-and-im-sharing.html#top1; I have
drawn on specific claims by Hillary and Autistic Chick in these posts. For other
self-advocacy bloggers whose advocacy despite their "low-functioning" status can be
(and has been) seen to challenge the perspective of Autism Speaks, see Larry Bissonnette and Tracy Thresher, *Wretches and Jabberers* (n.d.), http://wretchesandjabberers
.org; *Emma's Messiah Miracle of Music* (n.d.), http://emmasmiraclemusic.blogspot
.ca/; and Amy Sequenzia, *Non-speaking Autistic Speaking* (n.d.), http://nonspeaking
autisticspeaking.blogspot.ca/. See also Elizabeth J. Grace, *Tiny Grace Notes (Ask
an Autistic)* (n.d.), http://tinygracenotes.blogspot.ca/; Melanie Yergeau, *Aspie Rhetor*
(n.d.), http://autistext.com/(formerlyhttp://aspierhetor.com/); and NeuroQueer,
http://neuroqueer.blogspot.ca/. All sites accessed December 9, 2016.

9. The first quotation from Garland-Thomson is from "The Case for Conserving
Disability," *Journal of Bioethical Inquiry* 9, no. 3 (2012): 339–355, at 341; the quota-
tion in the final paragraph is on page 352. For the quotation from Kafer, see Alison
Kafer, *Feminist, Queer, Crip* (Bloomington: Indiana University Press, 2013), 83. For
a collection of classic work on standpoint theory that connects directly with
discussions of objectivity and neutrality in epistemology, see Sandra Harding, ed.,
The Feminist Standpoint Theory Reader: Intellectual and Political Controversies (New
York: Routledge, 2004), especially Alison Wylie's "Why Standpoint Matters" in that
volume, 339–351; see also Wylie's "Feminist Philosophy of Science: Standpoint
Matters," *Proceedings and Addresses of the American Philosophical Association* 86, no. 2
(2012): 47–76.

8 Eugenics as Wrongful Accusation

1. For the film in which Ken Nelson's comments were made, see *The Sterilization of
Leilani Muir*, directed by Glynis Whiting for the National Film Board of Canada. For
Ken's more recent reflections on eugenics as part of a broader first-person narrative,
see his interview at the Our Stories module at EugenicsArchive.ca.

2. For discussion of the ritual sexual abuse cases, see Debbie Nathan and Michael
Snedeker, *Satan's Silence: Ritual Abuse and the Making of a Modern American Witch
Hunt* (Lincoln, NE: Author's Choice Press, an imprint of IUniverse.com, Inc., 2001),
originally published by Basic Books in 1995; and Dorothy Rabinowitz, *No Crueler
Tyrannies: Accusation, False Witness, and Other Terrors of Our Times* (New York: The
Free Press, 2003), both of which derive from journalistic coverage by the authors
beginning, respectively, in the mid-1980s and the mid-1990s. The cases are also covered in several documentary films, including Andrew Jarecki's *Capturing the Friedmans* (HBO Documentary Films, 2003), focused on a case in Great Neck, New York,
and the more recent *Witch Hunt* (Hard-Nac Movies, Your Half Media Group: Glendale, CA, 2009), narrated and produced by Sean Penn, focused on the Bakersfield,
Kern County, cases in California.

3. The concepts of collective hysteria and groupthink arose in the last half of the nineteenth century as part of the idea of collective or crowd psychology, though the term "groupthink" was coined only in 1952 by W. H. Whyte. For classic studies of groupthink, see Irving Janis, *Victims of Groupthink: A Psychological Study of Foreign-Policy Decisions and Fiascoes* (Boston: Houghton Mifflin, 1972), and the rev. 2nd ed. titled *Groupthink: Psychological Studies of Policy Decisions and Fiascoes* (Boston: Houghton Mifflin, 1982). For discussions of the collective psychology tradition, see Robert Nye, *The Origins of Crowd Psychology: Gustav LeBon and the Crisis of Mass Democracy in the Third Republic* (Beverly Hills, CA: Sage Publications, 1975); Jaap van Ginneken, *Crowds, Psychology, and Politics 1871–1899* (New York: Cambridge University Press, 1992); and my brief discussion in *Boundaries of the Mind* (New York: Cambridge University Press, 2004), 269–274. The phrase "the banality of evil" comes from Hannah Arendt's best-known work, *Eichmann in Jerusalem: A Report on the Banality of Evil* (New York: Penguin Books, 1963). On the sociological concept of moral panics, see Stanley Cohen, *Folk Devils and Moral Panics: The Creation of the Mods and the Rockers* (London: McGibben and Kee, 1972). For the introduction of the illusion of explanatory depth and the conceptual framework on explanation and cognition in which it was embedded, see Robert A. Wilson and Frank C. Keil, "The Shadows and Shallows of Explanation," *Minds and Machines* 8 (1998): 137–159, reprinted with modification in Frank C. Keil and Robert A. Wilson, eds, *Explanation and Cognition* (Cambridge, MA: MIT Press, 2000), 87–114. For Keil's subsequent experimental exploration of the illusion, see Leonid Rozenblit and Frank C. Keil, "The Misunderstood Limits of Folk Science: An Illusion of Explanatory Depth," *Cognitive Science* 26, no. 5 (2002): 521–562; and Candice M. Mills and Frank C. Keil, "Knowing the Limits of One's Understanding: The Development of an Awareness of an Illusion of Explanatory Depth," *Journal of Experimental Child Psychology* 87, no. 1 (2004): 1–32. For the recent popularization I cite, see Steven Sloman and Philip Fernbach, *The Knowledge Illusion: Why We Never Think Alone* (New York: Penguin Random House, 2017).

4. See Judith Lewis Herman, *Father-Daughter Incest* (Cambridge, MA: Harvard University Press, [1981] 2000), republished with a new afterword; and her best-selling *Trauma and Recovery* (New York: Basic Books, [1992] 1997), also republished with a new afterword. The short quotation from Herman is from page 1, and the extract from page 7, of the republished *Trauma and Recovery*. For the introduction of post-traumatic stress disorder as a psychiatric category, see the *Diagnostic and Statistical Manual for Mental Disorders*, 3rd edition (New York: American Psychiatric Association, 1980). For two interesting discussions of sexual abuse, trauma, and post-traumatic stress disorder that do not discuss the ritual sexual abuse cases, see Allan Young, *The Harmony of Illusions: Inventing Post-traumatic Stress Disorder* (Princeton, NJ: Princeton University Press, 1995); and Joseph E. Davis, *Accounts of Innocence: Sexual Abuse, Trauma, and the Self* (Chicago: University of Chicago Press, 2005).

5. The long quotation from Herman comes from her afterword to the republished *Father-Daughter Incest*, 240–241. For early work on the looping effect, see Ian Hacking, "Making Up People," in *Reconstructing Individualism: Autonomy, Individuality, and the Self in Western Thought*, ed. Thomas C. Heller, Morton Sosna, and David E. Wellbery (Stanford: Stanford University Press, 1986), 222–236; and, for its most commonly cited expression, see Ian Hacking, "The Looping Effects of Human Kinds," in *Causal Cognition: A Multidisciplinary Debate*, ed. Dan Sperber, David Premack, and Anne J. Premack (New York: Clarendon Press, 1995), 351–394. For discussion of the Wenatchee case, see Rabinowitz, *No Crueler Tyrannies* (see note 2), 96–122; and Kathrine Beck, "Wenatchee Witch Hunt: Child Sex Abuse Cases in Douglas and Chelan Counties," October 28, 2004; accessed December 14, 2016, http://www.historylink .org/File/7065.

9   Knowing Agency

1. For a sense of the focus of contemporary discussions in what I am calling traditional analytic epistemology, see Matthias Steeup, "Epistemology," *Stanford Encyclopedia of Philosophy* (Fall 2016 edition), ed. Edward N. Zalta, accessed December 18, 2016, https://plato.stanford.edu/archives/fall2016/entries/epistemology/. For collections of influential essays on feminist standpoint theory, see Sandra Harding, ed., *The Feminist Standpoint Theory Reader: Intellectual and Political Controversies* (New York: Routledge, 2004); a special issue of the journal *Signs: Journal of Women and Culture* 22, no. 2 (1997); and a collection of reflective papers in *Hypatia: A Journal of Feminist Philosophy* 24, no. 4 (2009). For an influential and important collection of essays using the resources of analytic metaphysics and the philosophy of language that embraces situated knowledge without committing to standpoint epistemology, see Sally Haslanger, *Resisting Reality: Social Construction and Social Critique* (New York: Oxford University Press, 2012).

2. For work on the epistemology of scientific knowledge that is informed by work on gender and science, see Helen Longino, *Contextual Empiricism* (Princeton, NJ: Princeton University Press, 1990), and *The Fate of Knowledge* (Princeton, NJ: Princeton University Press, 2002); and Miriam Solomon, *Social Empiricism* (Cambridge, MA: MIT Press, 2001). The works of Fricker and Paul to which I refer are Miranda Fricker, *Epistemic Injustice: Power and the Ethics of Knowing* (New York: Oxford University Press, 2007) and Laurie Paul, *Transformative Experience* (New York: Oxford University Press, 2015). Patrick Rysiew's review of contextualism can be found in his "Epistemic Contextualism," *Stanford Encyclopedia of Philosophy* (Winter 2016 edition), ed. Edward N. Zalta, accessed December 18, 2016, https://plato.stanford.edu/ archives/win2016/entries/contextualism-epistemology/; for Grasswick's corresponding discussion of standpoint theory, see Heidi Grasswick, "Feminist Social Epistemology," *Stanford Encyclopedia of Philosophy* (Winter 2016 edition), ed. Edward N. Zalta, accessed December 18, 2016, https://plato.stanford.edu/archives/win2016/

entries/feminist-social-epistemology/. For an overview of naturalistic philosophy of science that covers ventures into the politics of differential knowing, see Philip Kitcher, "The Naturalists Return," *Philosophical Review* 101 (1992): 53–114; and for exemplary work in this tradition that articulates standpoint theory, see Joseph Rouse, "Feminism and the Social Construction of Scientific Knowledge," in *Feminism, Science and the Philosophy of Science*, ed. Lynne Hankinson Nelson and James Nelson (Dordrecht: Kluwer Academic Publishers, 1996), 195–215; and Alison Wylie, "Feminist Philosophy of Science: Standpoint Matters," *Proceedings and Addresses of the American Philosophical Association* 86, no. 2 (2012): 47–76. The short quotation I give from Wiley is from page 63. For the original work in the social sciences on feminist standpoint theory, see Dorothy Smith, "Women's Perspective as a Radical Critique of Sociology," *Sociological Inquiry* 44 (1974): 7–13; Nancy Hartsock, "The Feminist Standpoint: Developing the Ground for a Specifically Feminist Historical Materialism," in *Discovering Reality: Feminist Perspectives on Epistemology, Metaphysics, Methodology, and the Philosophy of Science*, ed. Sandra Harding and Merrill Hintikka (Dordrecht: D. Reidel, 1983), 283–310; and Patricia Hill Collins, "Learning from the Outsider Within: The Sociological Significance of Black Feminist Thought," *Social Problems* 33, no. 6 (1986): S14–S32; for its original development within philosophy, see Alison Jaggar, *Feminist Politics and Human Nature* (Totowa, NJ: Rowman and Littlefield, 1983) and Sandra Harding, *The Science Question in Feminism* (Ithaca, NY: Cornell University Press, 1986).

3. The special issue of the journal that I refer to is *Metaphilosophy* 40, no. 3–4 (July 2009); the corresponding book is Eva Feder Kittay and Licia Carlson, eds., *Cognitive Disability and Its Challenge to Moral Philosophy* (New York: Wiley-Blackwell, 2010), which contains four additional essays. The conference from which the papers derive was held in September 2008 in New York City, and the series of thirteen blogposts and surrounding discussion of video excerpts from the conference by contributors to *What Sorts of People Should There Be?* were posted as the "Thinking in Action" series between December 2008 and February 2009; they are collected at https:// whatsortsofpeople.wordpress.com/2009/02/13/all-wrapped-up-complete-thinking -in-action-series (accessed December 18, 2016). Post contributors included Dick Sobsey, Marc Workman, Angie Harris, Julie Maybee, Ron Amundson, and Rob Wilson. The two quotations I provide from Eva Feder Kittay are from her contribution to the conference, "The Personal Is Philosophical Is Political: A Philosopher and Mother of a Cognitively Disabled Person Sends Notes from the Battlefield," originally published in *Metaphilosophy* 40 (2009): 606–627, at 608–609 and 611, respectively.

4. For Leilani Muir's autobiography, see her *A Whisper Past: Childless after Eugenic Sterilization in* Alberta (Victoria, BC: Friesen Press, 2014); for her video narrative with the Living Archives project, see her contribution to the Our Stories module, accessed December 18, 2016, http://eugenicsarchive.ca/discover/our-stories/leilani. For further contextualization of these narratives, see her contributions to the films *The*

Sterilization of Leilani Muir (Ottawa: National Film Board of Canada, 1996) and *Surviving Eugenics* (Vancouver, BC: Moving Images Distribution, 2015). For a selection of narratives from those parenting around disability, see the other contributions to the Our Stories module, particularly the lengthy story told, in two parts, by Velvet Martin about her daughter Samantha, accessed December 18, 2016, http:// eugenicsarchive.ca/discover/our-stories/velvet.

5. For recent discussion of the citizen science movement, see Sandra Harding, *Objectivity and Diversity: Another Logic of Scientific Research* (Chicago: University of Chicago Press, 2015), especially chapters 1–2; and David Hess, "Science in an Era of Globalization: Alternative Pathways," in *Alternative Pathways in Science and Industry: Activism, Innovation, and the Environment in an Era of Globalization* (Cambridge, MA: MIT Press, 2007), 43–68, reprinted in abridged form in *The Postcolonial Science and Technology Studies Reader*, ed. Sandra Harding (Durham: Duke University Press, 2011), 419–437. For articulation of standpoint theory in terms of what she calls the "inversion thesis," see Alison Wylie, "Feminist Philosophy of Science: Standpoint Matters," *Proceedings and Addresses of the American Philosophical Association* 86, no. 2 (2012): 47–76, where Wylie characterizes that thesis as the view "that certain kinds of epistemic advantage accrue to those who are otherwise (socially, materially) disadvantaged" (57). For my own earlier discussion of knowing agents from the margins, see Robert A. Wilson, "The Role of Oral History in Surviving a Eugenic Past," in *Beyond Testimony and Trauma: Oral History in the Aftermath of Mass Violence*, ed. Steven High (Vancouver: University of British Columbia Press, 2015), 119–138.

10   Eugenics Unbound

1. For the recent location of standpoint theory in broader work on political commitment and science, see Sandra Harding, *Objectivity and Diversity: Another Logic of Scientific Research* (Chicago: University of Chicago Press, 2015) and "After Mr. Nowhere: What Kind of Proper Self for a Scientist?," *Feminist Philosophy Quarterly* 1, no. 1 (2015): article 2, accessed December 20, 2016. http://ir.lib.uwo.ca/fpq/vol1/ iss1/2. For the attempted reinvigoration of standpoint theory, see Susan Hekman, "Truth and Method: Feminist Standpoint Theory Revisited," *Signs: Journal of Women and Culture* 22 (1997): 341–365, which appeared with replies from Patricia Hill Collins, Sandra Harding, Dorothy Smith, and Nancy Hartsock. For the state of play in feminist standpoint theory in the first decade of the twenty-first century, see Sandra Harding, ed., *The Feminist Standpoint Theory Reader* (New York: Routledge, 2004), which includes the Hekman paper and these four replies, and the subsequent reflective papers in *Hypatia* 24, no. 4 (2009): 189–237, as well as Kristin Intemann, "25 Years of Feminist Empiricism and Standpoint Theory: Where Are We Now?," *Hypatia* 25, no. 4 (2010): 778–796.

2. For the original papers of Smith, Collins, and Hartsock, see Dorothy Smith, "Women's Perspective as a Radical Critique of Sociology," *Sociological Inquiry* 44 (1974): 7–13; Patricia Hill Collins, "Learning from the Outsider Within: The Sociological Significance of Black Feminist Thought," *Social Problems* 33, no. 6 (1986): S14–S32, and her *Black Feminist Thought: Knowledge, Consciousness, and the Politics of Empowerment* (New York: Routledge, [1991] 2000); and Nancy Hartsock, "The Feminist Standpoint: Developing the Ground for a Specifically Feminist Historical Materialism," in *Discovering Reality: Feminist Perspectives on Epistemology, Metaphysics, Methodology, and the Philosophy of Science*, ed. Sandra Harding and Merrill Hintikka (Dordrecht: D. Reidel, 1983), 283–310. The "Whose standpoint?" question was originally raised by Sandra Harding in her *Whose Science? Whose Knowledge? Thinking from Women's Lives* (Ithaca, NY: Cornell University Press, 1991), 124. For discussion of the achievement thesis, see Elizabeth Anderson, "Feminist Epistemology and Philosophy of Science," *Stanford Encyclopedia of Philosophy* (2015), accessed December 20, 2016, https://plato.stanford.edu/entries/feminism-epistemology/; Alison Wylie, "Why Standpoint Matters," in *The Feminist Standpoint Theory Reader*, ed. Sandra Harding (New York: Routledge, 2004), 339–351, originally published in *Science and Other Cultures: Issues in Philosophies of Science and Technology*, ed. Robert Figueroa and Sandra Harding (New York: Routledge, 2003), 26–48; and Alison Wylie, "Feminist Philosophy of Science: Standpoint Matters," *Proceedings and Addresses of the American Philosophical Association* 86, no. 2 (2012): 47–76. For the classic discussion of the role of intellectuals in class oppression and revolution, see Georg Lukács, *History and Class Consciousness: Studies in Marxist Dialectics* (Cambridge, MA: MIT Press, [1923] 1971), originally published in German. For the application of standpoint theory to race, see the works cited by Collins above; for applications to disability, see Susan Wendell, *The Rejected Body: Feminist Philosophical Reflections on Disability* (New York: Routledge, 1996); and Mary B. Mahowald, "A Feminist Standpoint," in *Disability, Difference, Discrimination: Perspectives on Justice in Bioethics and Public Policy*, ed. Anita Silvers, David T. Wasserman, and Mary B. Mahowald (Lanham, MD: Rowman & Littlefield, 1998), 209–252.

3. The full reference for *Black Feminist Thought* is given in note 2, and all short quotations from Collins in section 10.3 are taken from pages 26 and 28 of that source; the long, offset quotation and surrounding discussion can be found on page 270. For Collins's exchange with Hekman, see the work cited by Hekman in the note 2, and Patricia Hill Collins, "Comment on Hekman's 'Truth and Method: Feminist Standpoint Theory Revisited': Where's the Power?," *Signs: Journal of Women and Culture* 22 (1997): 375–381, also reprinted in Sandra Harding's anthology on feminist standpoint theory cited in note 1.

4. For representative, recent work on collective intentionality, see Bryce Huebner, *Macrocognition: A Theory of Distributed Minds and Collective Intentionality* (New York: Oxford University Press, 2014); Marija Jankovic and Kirk Ludwig, eds., *Routledge Handbook of Collective Intentionality* (New York: Routledge, 2017); and Michelle L.

Meade, Celia B. Harris, Penny Van Bergen, John Sutton, and Amanda J. Barnier, eds., *Collaborative Remembering: Theories, Research, and Applications* (New York: Oxford University Press, 2017). For the idea of extended cognition, see my *Boundaries of the Mind* (New York: Cambridge University Press, 2004); and Andy Clark, *Supersizing the Mind: Embodiment, Action, and Cognitive Extension* (New York: Oxford University Press, 2008). For an interesting, recent departure from traditional individualistic views of autonomy in a discussion of intellectual disability, see Laura Davey, "Philosophical Inclusive Design: Intellectual Disability and the Limits of Individual Autonomy in Moral and Political Theory," *Hypatia* 30, no. 1 (2015): 132–148. For relevant deployments of the notion of *mitgefuhl*, see Alexis Shotwell, *Knowing Otherwise: Race, Gender, and Implicit Understanding* (University Park: Pennsylvania State University Press, 2011); and Barbara Weber, *Zwischen Vernunft und Mitgefühl* (Freiburg: Alber Publishers, 2013).

5. For a focused, succinct discussion of sorts of people, see my "Sorts of People," April 29, 2014, accessed December 20, 2016, http://eugenicsarchive.ca/discover/enc yclopedia/535eee527095aa000000025c. For one of the few book-length philosophical treatments of intellectual disability, see Licia Carlson, *The Faces of Intellectual Disability: Philosophical Reflections* (Bloomington: Indiana University Press, 2009); and for a recent paper on philosophical theorizing and intellectual disability, see Ashley Taylor, "Expressions of 'Lives Worth Living' and Their Foreclosure through Philosophical Theorizing on Moral Status and Intellectual Disability," in *Foucault and the Government of Disability*, rev. ed., ed. Shelley Tremain (Ann Arbor: University of Michigan Press, 2015), 372–395.

6. Individual survivor stories are available at the Our Stories module at Eugenics Archive.ca, accessed December 20, 2016, http://eugenicsarchive.ca/discover/our -stories, and are the basis for the documentary film, *Surviving Eugenics* (2015), http:// eugenicsarchive.ca/film, distributed by Moving Images Distribution in Vancouver, British Columbia. For a sample of the published record of psychotropic experimentation on children at Red Deer's Provincial Training School by its medical superintendent, see Leonard J. Le Vann, "Trifluoperazine Dihydrochloride: An Effective Tranquillizing Agent for Behavioural Abnormalities in Defective Children," *Canadian Medical Association Journal* 80, no. 2 (1959): 123–124; "Thioridazine (Mellaril) A Psycho-sedative Virtually Free of Side-effects," *Alberta Medical Bulletin* 26 (1961): 144–147; "A New Butyrophenone: Trifluperidol: A Psychiatric Evaluation in a Pediatric Setting," *The Canadian Psychiatric Association Journal* 13 (1968): 271–273; "Haloperidol in the Treatment of Behavioral Disorders in Children and Adolescents," *Canadian Psychiatric Association Journal* 14 (1969): 217–220; "Clinical Comparison of Haloperidol and Chlorpromazine in Mentally Retarded Children," *American Journal of Mental Deficiency* 75 (1971): 719–723. For the medical experimentation on boys diagnosed with Down syndrome who were resident at the Provincial Training School, see the account of the testimony of Alberta Eugenics Board member Margaret Thompson in paragraph 114 of the summary of Muir v. Alberta, 123 *Dominion*

Law Reports (4th series) (1996), 695–762, 1996 CanLII 7287 (QB AB), as well as the account that Sandra Anderson gives of her cross-examination of Thompson in my 2013 interview with Anderson at EugenicsArchive.ca, accessed October 23, 2016, http://eugenicsarchive.ca/discover/interviews. See especially 19.45–22.15 of the interview, with preceding discussion of Le Vann and following discussion of Thompson. An excerpt from this part of the interview is included in the film *Surviving Eugenics*.

7. For the Our Stories module, see note 6. For the representative quotations from Roy Skoreyko, Glenn Sinclair, and Kyle Lillo, see, respectively, 2.00–3.54 and 5.58–6.19 of Roy's story, 1.00–1.48 of Glenn's story, and 1.40–2.14 of Kyle's story. Kyle and his mother, Lorna Lillo, have also recorded a joint narrative available at the Our Stories module that provides broader background to Kyle's life: accessed December 20, 2016, http://eugenicsarchive.ca/discover/our-stories/kyle-lorna. For earlier, non-video narratives of Albertans with developmental disabilities who were institutionalized, including those at the Provincial Training School in Red Deer, see *Hear My Voice: Stories Told by Albertans with Developmental Disabilities Who Were Once Institutionalized* (Edmonton: Alberta Association for Community Living, 2006).

References

Adams, Frederick, and Ken Aizawa. 2008. *The Bounds of Cognition.* New York: Blackwell.

Agar, Nicholas. 2004. *Liberal Eugenics: In Defence of Human Enhancement.* Malden, MA: Blackwell.

Agar, Nicholas. 2010. *Humanity's End: Why We Should Reject Radical Enhancement.* Cambridge, MA: MIT Press.

Agar, Nicholas. 2013. "Eugenics, Old and Neoliberal Theories of." In *Encyclopedia of Philosophy and the Social Sciences,* ed. Byron Kaldis, 283–289. Thousand Oaks, CA: Sage Publications.

Alberta Association for Community Living. 2006. *Hear My Voice: Stories Told by Albertans with Developmental Disabilities Who Were Once Institutionalized.* Edmonton: Alberta Association for Community Living.

Aldrich, Morton Arnold, William Herbert Carruth, Charles Benedict Davenport, Charles Abram Ellwood, Arthur Holmes, William Henry Howell, Harvey Ernest Jordan et al. 1914. *Eugenics: Twelve University Lectures.* New York: Dodd, Mead.

Allen, Garland. 1986. "The Eugenics Record Office at Cold Spring Harbor, 1910–1940: An Essay in Institutional History." *Osiris* 2:225–264.

Ambler, Jacalyn. 2014. "Immigration." Encyclopedia module, EugenicsArchives.ca. http://eugenicsarchive.ca/discover/encyclopedia/53480910132156674b0002b6.

American Psychiatric Association. 1980. *Diagnostic and Statistical Manual for Mental Disorders.* 3rd ed. New York: American Psychiatric Association.

Anderson, Elizabeth. 2015. "Feminist Epistemology and Philosophy of Science." *Stanford Encyclopedia of Philosophy.* https://plato.stanford.edu/entries/feminism-epistemology/.

Archibald, J. David. 2014. *Aristotle's Ladder, Darwin's Tree: The Evolution of Visual Metaphors for Biological Order.* New York: Columbia University Press.

Arendt, Hannah. 1963. *Eichmann in Jerusalem: A Report on the Banality of Evil*. New York: Penguin Books.

Aristotle. 1984. *The Complete Works of Aristotle: The Revised Oxford Translation*, vol. 2. Ed. Jonathan Barnes. Princeton, NJ: Princeton University Press.

Armer, Bill. 2007. "Eugenetics: A Polemical View of Social Policy in the Genetic Age." *New Formations* 60:89–101.

Asch, Adrienne. 1989. "Reproductive Technology and Disability." In *Reproductive Laws for the 1990s*, ed. Sherrill Cohen and Nadine Taub, 69–124. Clifton, NJ: Humana Press.

Asch, Adrienne. 2003. "Disability Equality and Prenatal Testing: Contradictory or Compatible?" *Florida State University Law Review* 30 (2): 315–342.

Asch, Adrienne, and Michelle Fine, eds. 1988. *Women with Disabilities: Essays in Psychology, Culture, and Politics*. Philadelphia, PA: Temple University Press.

Asch, Adrienne, and David Wasserman. 2005. "Where Is the Sin in Synecdoche?: Prenatal Testing and the Parent-Child Relationship." In *Quality of Life and Human Difference: Genetic Testing, Health Care, and Disability*, ed. David Wasserman, Jerome Bickenbach, and Robert Wachbroit, 172–216. New York: Cambridge University Press.

Autism Speaks. 2010. "What Is Asperger Syndrome/HFA?" https://www.autismspeaks .org/family-services/tool-kits/asperger-syndrome-and-high-functioning-autism-tool -kit/introduction.

Autistic Advocacy. 2016. "Before You Donate to Autism Speaks, Consider the Facts." http://autisticadvocacy.org/wp-content/uploads/2016/03/AutismSpeaksFlyer_color _2016.pdf.

Autistic Chick. 2014. "Nonspeaking. Real. Self-Advocates." http://autisticchick .blogspot.ca/2014/08/i-left-comment-on-thing-and-im-sharing.html#top1.

Ball, Natalie. 2014. "Le Vann, Leonard J." Players module, EugenicsArchives.ca. http://eugenicsarchive.ca/discover/players/512fa4b134c5399e2c00000d.

Bamford, Sandra, and James Leach, eds. 2009. *Kinship and Beyond: The Genealogical Model Reconsidered*. New York: Berghann Press.

Barker, Matthew J., and Robert A. Wilson. Unpublished manuscript. "Well-Being, Disability, and Choosing Children."

Barr, Martin W. 1904. *Mental Defectives: Their History, Treatment and Training*. Philadelphia: P. Blakiston's.

Barr, Martin W. 1920. "Some Notes on Asexuality; With a Report of Eighteen Cases." *Journal of Nervous and Mental Disease* 51 (3): 231–241.

Bashford, Alison, and Philippa Levine, eds. 2010. *The Oxford Handbook of the History of Eugenics*. New York: Oxford University Press.

Baynton, Douglas C. 2016. *Defectives in the Land: Disability and Immigration in the Age of Eugenics*. Chicago: University of Chicago Press.

BBC News. 2014. "India's Dark History of Sterilisation." November 14. http://www .bbc.com/news/world-asia-india-30040790.

Beck, Kathrine. 2004. "Wenatchee Witch Hunt: Child Sex Abuse Cases in Douglas and Chelan Counties." October 28. http://www.historylink.org/File/7065.

Begos, Kevin, Danielle Deaver, John Railey, and Scott Sexton. 2012. *Against Their Will: North Carolina's Sterilization Program and the Campaign for Reparations*. Florida: Gray Oak Books.

Ben-Moshe, Liat, Chris Chapman, and Allison C. Carey, eds. 2014. *Disability Incarcerated: Imprisonment and Disability in the United States and Canada*. New York: Palgrave Macmillan.

Benatar, David, and David Wasserman. 2015. *Debating Procreation: Is It Wrong to Reproduce?* New York: Oxford University Press.

Bérubé, Michael. 2014. "Down Syndrome." Encyclopedia module, EugenicsArchives.ca. http://www.eugenicsarchive.ca/discover/encyclopedia/535eeb507095aa000000021d.

Billinger, Michael. 2014. "Ethnicity and Race." Connections module, EugenicsArchives.ca. http://eugenicsarchive.ca/discover/connections/5508686c5f8ecc9f71000003.

Billinger, Michael. 2014. "Miscegenation." Encyclopedia module, EugenicsArchives.ca. http://eugenicsarchive.ca/discover/encyclopedia/52329c0e5c2ec5000000000b.

Binding, Karl, and Alfred Hoche. 1920. *Die Freigabe der Vernichtung Lebensunwerten Lebens: Ihr Mass und Ihr Ziel* (Leipzig: Felix Meiner). Second edition available at http://www.gutenberg.org/ebooks/44565 and from http://chillingeffects.de/binding .pdf. Translated by Walter E. Wright as "Permitting the Destruction of Unworthy Life: Its Extent and Form" in 1992, in *Issues in Law and Medicine* 8 (2): 231–265. Translated with a commentary in 1975 as *The Release of the Destruction of Life Devoid of Value: It's Measure and It's Form* [sic]. Santa Ana, CA: R. L. Sasson.

Bissonnette, Larry, and Tracy Thresher. n.d. *Wretches and Jabberers*. http:// wretchesandjabberers.org.

Blackmar, Frank W. 1897. "The Smoky Pilgrims." *American Journal of Sociology* 2 (4): 485–500.

Boring, Edwin G. 1929. *A History of Experimental Psychology*. New York: Century.

Bouquet, Mary. 1996. "Family Trees and Their Affinities: The Visual Imperative of the Genealogical Method." *Man* (n.s.) 2 (1): 43–66.

Boyd, P., C. DeVigan, B. Khoshnood, M. Loane, E. Garne, H. Dolk, and the EURO-CAT working group. 2008. "Survey of Prenatal Screening Policies in Europe for Structural Malformations and Chromosome Anomalies, and Their Impact on Detection and Termination Rates for Neural Tube Defects and Down's Syndrome." *BJOG: An International Journal of Obstetrics and Gynaecology* 115:689–696.

Broberg, Gunnar, and Nils Roll-Hansen, eds. 1996. *Eugenics and the Welfare State: Sterilization Policy in Denmark, Sweden, Norway, and Finland.* East Lansing: Michigan State University Press.

Brown, R., J. Taylor, and B. Matthews. 2001. "Quality of Life—Ageing and Down Syndrome." *Down's Syndrome: Research and Practice* 6 (3): 111–116.

Brownstein, Michael. 2016. "Implicit Bias." *Stanford Encyclopedia of Philosophy* (Spring edition), ed. Edward N. Zalta. https://plato.stanford.edu/archives/spr2016/entries/implicit-bias/.

Brownstein, Michael, and Jennifer Saul, eds. 2016. *Implicit Bias and Philosophy: Volume 2, Moral Responsibility, Structural Injustice, and Ethics.* Oxford: Oxford University Press.

Bruinius, Harry. 2006. *Better for All the World: The Secret History of Forced Sterilization and America's Quest for Racial Purity.* New York: Alfred J. Knopf.

Buchanan, Alan, Dan Brock, Norman Daniels, and Dan Wikler. 2000. *From Chance to Choice: Genetics and Justice.* Cambridge: Cambridge University Press.

Bulmer, Michael G. 2003. *Francis Galton: Pioneer of Heredity and Biometry.* Baltimore, MD: Johns Hopkins University Press.

Butler, Judith. 2009. *Frames of War: When Is Life Grievable?* New York: Verso.

Byrne, Richard, and Andrew Whiten, eds. 1989. *Machiavellian Intelligence: Social Expertise and the Evolution of Intellect in Monkeys, Apes, and Humans.* New York: Oxford University Press.

Byrne, Richard, and Andrew Whiten, eds. 1997. *Machiavellian Intelligence II: Extensions and Evaluations.* New York: Oxford University Press.

The Caffeinated Autistic. 2014. "New Autism Speaks Masterpost." https://thecaffeinatedautistic.wordpress.com/new-autism-speaks-masterpost-updated-62014/.

California Senate Resolution 20. 2003. http://www.csus.edu/cshpe/eugenics/docs/senate_resolution_20.pdf.

Canadian Association for Community Living. 2016. "Project Value: Disabled Lives Have Value." http://www.cacl.ca/news-stories/blog/project-value-disabled-lives-have-value.

Canadian Bulletin of Medical History 31 (1) (2014). Special issue on "Eugenics," ed. Erika Dyck.

Canguilhem, George. [1966] 1978. *On the Normal and the Pathological*. Dordrecht: D. Reidel.

Carlson, Elof Axel. 2001. *The Unfit: A History of a Bad Idea*. New York: Cold Spring Harbor Laboratory Press.

Carlson, Licia. 2009. *The Faces of Intellectual Disability: Philosophical Reflections*. Bloomington: Indiana University Press.

CBC Radio. "Aboriginal Women Say They Were Sterilized against Their Will." *The Current*, January 7, 2016. http://www.cbc.ca/radio/thecurrent/the-current-for -january-7-2016-1.3393099/aboriginal-women-say-they-were-sterilized-against-their -will-in-hospital-1.3393143.

Charleton, James I. 1998. *Nothing about Us without Us: Disability Oppression and Empowerment*. Berkeley: University of California Press.

Clare, Eli. 2009. *Exile and Pride: Disability, Queerness, and Liberation*. Durham, NC: Duke University Press.

Clark, Andy. 2008. *Supersizing the Mind: Embodiment, Action, and Cognitive Extension*. New York: Oxford University Press.

CNN. 2014. "India Sterilization Program under Fire after Women's Deaths." November 12. http://www.cnn.com/2014/11/12/world/asia/india-sterilization-deaths.

Cohen, Adam. 2016. *Imbeciles: The Supreme Court, American Eugenics, and the Sterilization of Carrie Buck*. New York: Penguin USA.

Cohen, Stanley. 1972. *Folk Devils and Moral Panics: The Creation of the Mods and the Rockers*. London: McGibben and Kee.

Collins, Patricia Hill. 1986. "Learning from the Outsider Within: The Sociological Significance of Black Feminist Thought." *Social Problems* 33 (6): S14–S32.

Collins, Patricia Hill. 1997. "Comment on Hekman's 'Truth and Method: Feminist Standpoint Theory Revisited': Where's the Power?" *Signs: Journal of Women and Culture* 22:375–381.

Collins, Patricia Hill. [1991] 2000. *Black Feminist Thought: Knowledge, Consciousness, and the Politics of Empowerment*. New York: Routledge.

Cullen, David. 2007. "Back to the Future: Eugenics—A Bibliographic Essay." *Public Historian* 29 (3): 163–175.

Danielson, Florence H., and Charles B. Davenport. 1912. *The Hill Folk: Report on a Rural Community of Hereditary Defectives*. Memoir no. 1. Cold Spring Harbor, NY: Eugenics Record Office [Lancaster, PA: Press of the New Era Printing Company].

Darwin, Charles. 1859. *On the Origin of Species by Means of Natural Selection, or the Preservation of Favoured Races in the Struggle for Life*. London: Murray.

Davenport, Charles B. 1912. *The Family History Book*. Eugenics Record Office Bulletin No. 7. Cold Spring Harbor, NY: Eugenics Record Office.

Davenport, Charles B. 1912. "The Inheritance of Physical and Mental Traits of Man and Their Application to Eugenics." In *Heredity and Eugenics*, ed. William E. Castle, John M. Coulter, Charles B. Davenport, Edward M. East and William L. Tower, 269–288. Chicago: University of Chicago Press.

Davenport, Charles B. 1912. *The Trait Book*. Eugenics Record Office Bulletin No. 6. Cold Spring Harbor, NY: Eugenics Record Office.

Davenport, Charles. 1914. "The Eugenics Programme and Progress in Its Achievement." In *Eugenics: Twelve University Lectures*, Morton Arnold Aldrich, William Herbert Carruth, Charles Benedict Davenport, Charles Abram Ellwood, Arthur Holmes, William Henry Howell, Harvey Ernest Jordan et al., 1–14. New York: Dodd, Mead.

Davenport, Charles B. 1915. "Huntington's Chorea in Relation to Heredity and Eugenics." *Proceedings of the National Academy of Sciences of the United States of America* 1:283–285.

Davenport, Charles B., and Harry H. Laughlin, D. F. Weeks, E. R. Johnstone, and H. H. Goddard. 1911. *The Study of Human Heredity: Methods of Collecting, Charting and Analyzing Data*. Eugenics Record Office Bulletin No. 2. Cold Spring Harbor, NY: Eugenics Record Office.

Davenport, Charles B., and Elizabeth B. Muncey. 1916. "Huntington's Chorea in Relation to Heredity and Eugenics." *American Journal of Insanity [Psychiatry]* 73:195–222. Reprinted as Eugenics Record Office Bulletin No. 17.

Davenport, Charles B. and David Fairchild Weeks. 1911. "A First Study of Inheritance of Epilepsy." *Journal of Nervous and Mental Disease* 38: 641–670. Reprinted as Eugenics Record Office Bulletin No. 4.

Davenport, Gertrude C. 1907. "Hereditary Crime." *American Journal of Sociology* 13 (3): 402–409.

Davey, Laura. 2015. "Philosophical Inclusive Design: Intellectual Disability and the Limits of Individual Autonomy in Moral and Political Theory." *Hypatia* 30 (1): 132–148.

Davis, Angela. 2014. "On Prisons, Race, and Gender Based Violence." March 15, Lecture at the University of Alberta, Edmonton, Canada.

Davis, Joseph E. 2005. *Accounts of Innocence: Sexual Abuse, Trauma, and the Self*. Chicago: University of Chicago Press.

Davis, Lennard. 1995. *Enforcing Normalcy: Deafness, Disability and the Body*. London: Verso.

Davis, Lennard. 2013. *The End of Normalcy: Identity in a Biocultural Era*. Lansing: University of Michigan Press.

Davis, Lennard. [1995] 2016. *The Disability Studies Reader*. 5th ed. New York: Routledge.

Demazeux, Steeves. 2014. "Psychiatric Classification." Connections module, EugenicsArchives.ca. http://eugenicsarchive.ca/discover/connections/53d831744 c879d000000000f.

Disability Rights Education and Defense Fund. 2007. "Investigative Report Regarding the 'Ashley Treatment.'" May. https://dredf.org/public-policy/ethics/ investigative-report-regarding-the-ashley-treatment/.

Dowbiggin, Ian. 1995. "'Keeping This Young Country Sane': C. K. Clarke, Immigration Restriction, and Canadian Psychiatry, 1890–1925." *Canadian Historical Review* 76:598–627.

Dowbiggin, Ian R. 2003. *Keeping America Sane: Psychiatry and Eugenics in the United States and Canada, 1880–1940*. Ithaca: Cornell University Press.

Dugdale, Richard A. 1877. *The Jukes: A Study in Crime, Pauperism, Disease, and Heredity*. Boston: Putnam.

Durkheim, Emile. 1974. "Individual and Collective Representations." In *Sociology and Philosophy*, 1–34. New York: Simon and Schuster.

Duster, Troy. [1990] 2003. *Backdoor to Eugenics*. New York: Routledge.

Dyck, Erika. 2013. *Facing Eugenics: Reproduction, Sterilization, and the Politics of Choice*. Toronto: University of Toronto Press.

Ebbinghaus, Hermann. 1908. *Psychology: An Elementary Textbook*. Boston: Heath.

Edwards, Martha L. 1997. "Constructions of Physical Disability in the Ancient World—The Community Concept." In *The Body and Physical Difference: Discourses of Disability*, ed. David T. Mitchell and Sharon L. Snyder, 35–50. Ann Arbor: University of Michigan Press.

Edwards, Martha·L. 1997. "Deaf and Dumb in Ancient Greece." In *The Disability Studies Reader*, 2nd ed., ed. Lennard Davis, 29–51. New York: Routledge.

Ehrenreich, Barbara, and John Ehrenreich. 1971. *The American Health Empire: Power, Profits, and Politics*. New York: Vintage.

Emma's Messiah Miracle of Music. n.d. http://emmasmiraclemusic.blogspot.ca/.

Estabrook, Arthur H. 1916. *The Jukes in 1915*. Washington, DC: The Carnegie Institution.

Estabrook, Arthur H., and Charles B. Davenport. 1912. *The Nam Family: A Study in Cacogenics*. Cold Spring Harbor, NY: Eugenics Record Office [Lancaster, PA: Press of the New Era Printing Company].

Estabrook, Arthur H., and Ivan E. McDougle. 1926. *Mongrel Virginians: The Win Tribe*. Baltimore: Williams and Wilkins.

Estes, Carroll L., Charlene Harrington, and David N. Pellow. 2000. "Medical-Industrial Complex." In *Encyclopedia of Sociology*, 2nd ed., vol. 3, ed. Edgar F. Borgatta and Rhonda J. V. Montgomery, 1818–1832. New York: Macmillan.

Evans, Meg. 2009. "Why Autism Speaks Does Not Speak for Us." Autistic Self-Advocacy Network, Central Ohio. http://asancentralohio.blogspot.ca/2009/08/why -autism-speaks-does-not-speak-for-us.html.

Ewin, Robert E. 1981. *Cooperation and Human Values: A Study in Moral Reasoning*. New York: St. Martin's Press.

Ewin, Robert E. 1991. *Virtues and Rights: The Moral Philosophy of Thomas Hobbes*. Boulder, CO: Westview Press.

Facing History and Ourselves. 2002. *Race and Membership in American History: The Eugenics Movement*. Brookline, MA: Facing History and Ourselves Foundation.

Finlayson, Anna Wendt. 1916. *The Dack Family: A Study in Hereditary Lack of Emotional Control*. Eugenics Records Office Bulletin no. 15. Cold Spring Harbor, NY: Eugenics Record Office.

Fisher, Ronald. 1930. *The Genetical Theory of Natural Selection*. Oxford: Clarendon Press.

Foucault, Michel. 2003. *Abnormal: Lectures at the College de France 1974–75*. Trans. G. Burchell. New York: Picador.

Foucault, Michel. 1965. *Madness and Civilization*. Trans. R. Howard. New York: Random House.

Foucault, Michel. 1979. *Discipline and Punish: The Birth of the Prison*. Trans. A. Sheridan. New York: Vintage.

Foucault, Michel. 2003. *"Society Must Be Defended": Lectures at the College de France 1975–76*. Trans. D. Macey. New York: Picador.

Foucault, Michel. 2006. *History of Madness*. London: Routledge.

Fournier, Marcel. 2005. "Durkheim's Life and Context: Something New About Durkheim?" In *The Cambridge Companion to Durkheim*, ed. Jeffrey C. Alexander and Philip Smith, 41–69. New York: Cambridge University Press.

Freeman, Samuel. 2014. "Original Position." *Stanford Encyclopedia of Philosophy*. http://plato.stanford.edu/entries/original-position/.

Fricker, Miranda. 2007. *Epistemic Injustice: Power and the Ethics of Knowing*. New York: Oxford University Press.

Galton, Francis. 1874. *English Men of Science: Their Nature and Nurture*. London: Macmillan.

Galton, Francis. 1865. "Hereditary Talent and Character." *Macmillan's Magazine* 12:157–166, 318–327.

Galton, Francis. 1869. *Hereditary Genius : An Inquiry into Its Laws and Consequences*. London: MacMillan.

Galton, Francis. 1883. *Inquiries into Human Faculty and Its Development*. London: MacMillan.

Galton, Francis. 1908. "Race Improvement." In *Memories of My Life*, 310–323. London: Methuen.

Garland-Thomson, Rosemarie. 2012. "The Case for Conserving Disability." *Journal of Bioethical Inquiry* 9 (3): 339–355.

Gilbert, Margaret. 1996. *Living Together: Rationality, Sociality, and Obligation*. New York: Rowman and Littlefield.

Gillham, Nicholas W. 2001. *A Life of Sir Francis Galton: From African Exploration to the Birth of Eugenics*. Oxford: Oxford University Press.

Gil-White, Francisco. 2001. "Are Ethnic Groups Biological 'Species' to the Human Brain?" *Current Anthropology* 42 (4): 515–553.

Glover, Jonathan. 1984. *What Sort of People Should There Be? Genetic Engineering, Brain Control, and Their Impact on Our Future World*. New York: Penguin Books.

Glover, Jonathan. 2006. *Choosing Children: Genes, Disability, and Design*. Oxford: Clarendon Press.

Goddard, Henry H. 1912. *The Kallikak Family: A Study in the Heredity of Feeble-Mindedness*. New York: Macmillan.

Goddard, Henry H. 1914. *Feeble-Mindedness: Its Causes and Consequences*. New York: Macmillan.

Godfrey-Smith, Peter. 2016. *Other Minds: The Octopus, the Sea, and the Deep Origins of Consciousness*. New York: Farrar, Strauss, and Giroux.

Goffman, Erving. 1961. *Asylums: Essays on the Social Situation of Mental Patients and Other Inmates*. New York: Anchor Books.

Gothard, Jan. 2011. *Greater Expectations: Living with Down Syndrome in the 21st Century*. Fremantle: Fremantle Press.

Government of Canada. 1923. "An Act Respecting Chinese Immigration." *Acts of the Parliament of the Dominion of Canada*. Ottawa: Brown Chamberlin, Law Printer (for Canada) to the Queen's Most Excellent Majesty, 1873–1951. 20 v.; 23–26 cm. Chapter 32, section 8. http://www.collectionscanada.gc.ca/immigrants/021017 -2412.02-e.html.

Government of Canada. 2016. Bill C-14. http://www.parl.gc.ca/HousePublications/ Publication.aspx?Language=E&Mode=1&DocId=8384014; summary at https:// openparliament.ca/bills/42-1/C-14/.

Grace, Elizabeth J. n.d. *Tiny Grace Notes (Ask an Autistic)*. http://tinygracenotes .blogspot.ca/.

Grasswick, Heidi. 2016. "Feminist Social Epistemology." *Stanford Encyclopedia of Philosophy* (Winter edition), ed. Edward N. Zalta. https://plato.stanford.edu/archives/ win2016/entries/feminist-social-epistemology/.

Greenberg, Daniel S. [1967] 1999. *The Politics of Pure Science*. Chicago: University of Chicago Press.

Greenwood, Jennifer. 2015. *Becoming Human: The Ontogenesis, Metaphysics, and Expression of Human Emotionality*. Cambridge, MA: MIT Press.

Grekul, Jana, Harvey Krahn, and David Odynak. 2004. "Sterilizing the 'Feeble-Minded': Eugenics in Alberta, Canada, 1929–1972." *Journal of Historical Sociology* 17 (4): 358–384.

The Guardian. 2012. "Growth Attenuation Treatment: Tom, the First Boy to Undergo Procedure." March 16. http://www.guardian.co.uk/society/2012/mar/16/ growth-attenuation-treatment-toms-story.

The Guardian. 2014. "India Mass Sterilisation: Women Were 'Forced' into Camps, Say Relatives." November 12. https://www.theguardian.com/world/2014/nov/12/ india-sterilisation-deaths-women-forced-camps-relatives.

Gunther, Daniel F., and Douglas S. Diekema. 2006. "Attenuating Growth in Children with Profound Developmental Disability: A New Approach to an Old Dilemma." *Archives of Pediatrics & Adolescent Medicine* 160 (10): 1013–1017.

Haller, Mark. 1963. *Hereditarian Attitudes in American Thought*. New Brunswick: Rutgers University Press.

Hanes, Roy. n.d. "None Is Still Too Many: An Historical Exploration of Canadian Immigration Legislation as It Pertains to People with Disabilities." Council of Canadians with Disabilities. http://www.ccdonline.ca/en/socialpolicy/access-inclusion/ none-still-too-many.

Hansen, Randall, and Desmond King. 2013. *Sterilized by the State: Eugenics, Race, and the Population Scare in Twentieth-Century North America*. New York: Cambridge University Press.

Haraway, Donna. 1991. "A Cyborg Manifesto: Science, Technology, and Socialist-Feminism in the Late Twentieth Century." In *Simians, Cyborgs and Women: The Reinvention of Nature*, 149–181. New York: Routledge.

Harding, Sandra. 1986. *The Science Question in Feminism*. Ithaca, NY: Cornell University Press.

Harding, Sandra. 1991. *Whose Science? Whose Knowledge? Thinking from Women's Lives*. Ithaca, NY: Cornell University Press.

Harding, Sandra, ed. 2004. *The Feminist Standpoint Theory Reader: Intellectual and Political Controversies*. New York: Routledge.

Harding, Sandra. 2015. "After Mr. Nowhere: What Kind of Proper Self for a Scientist?" *Feminist Philosophy Quarterly* 1 (1): article 2. http://ir.lib.uwo.ca/fpq/vol1/iss1/2.

Harding, Sandra. 2015. *Objectivity and Diversity: Another Logic of Scientific Research*. Chicago: University of Chicago Press.

Harris-Zsovan, Jane. 2010. *Eugenics and the Firewall: Canada's Nasty Little Secret*. Winnipeg, MB: J. Gordon Shillingford Publishing.

Hartsock, Nancy. 1983. "The Feminist Standpoint: Developing the Ground for a Specifically Feminist Historical Materialism." In *Discovering Reality: Feminist Perspectives on Epistemology, Metaphysics, Methodology, and the Philosophy of Science*, ed. Sandra Harding and Merrill Hintikka, 283–310. Dordrecht: D. Reidel.

Haslam, Nick, and Stephen Loughnan. 2014. "Dehumanization and Infrahumanization." *Annual Review of Psychology* 65:399–423.

Haslanger, Sally. 2012. *Resisting Reality: Social Construction and Social Critique*. New York: Oxford University Press.

Hekman, Susan. 1997. "Truth and Method: Feminist Standpoint Theory Revisited." *Signs: Journal of Women and Culture* 22:341–365.

Herissone-Kelly, P. 2006. "Procreative Beneficence and the Prospective Parent." *Journal of Medical Ethics* 32 (3): 166–169.

Herman, Judith Lewis. [1981] 2000. *Father-Daughter Incest*. Cambridge, MA: Harvard University Press.

Herman, Judith Lewis. [1992] 1997. *Trauma and Recovery*. New York: Basic Books.

Hess, David. 2007. "Science in an Era of Globalization: Alternative Pathways." In *Alternative Pathways in Science and Industry: Activism, Innovation, and the Environment in an Era of Globalization*, 43–68. Cambridge, MA: MIT Press. Reprinted in abridged form in *The Postcolonial Science and Technology Studies Reader*, ed. Sandra Harding, 419–437. Durham: Duke University Press, 2011.

Hillary, Alyssa. 2014. "'Autism-Speaks Are Work-Stealing, White-Texting Liars,' Yes, That Too." http://yesthattoo.blogspot.ca/2014/01/autism-speaks-are-work-stealing -white.html.

Himmelstein, David, Deborah Thorne, Elizabeth Warren, and Steffie Woolhandler. 2009. "Medical Bankruptcy in the United States, 2007: Results of a National Study." *American Journal of Medicine* 122:741–746.

Hobbes, Thomas. [1651] 1991. *Leviathan*. Cambridge: Cambridge University Press.

Hubbard, Ruth. 1984. "Personal Courage Is Not Enough: Some Hazards of Child-bearing in the 1980s." In *Test-Tube Women: What Future for Motherhood*, ed. Rita Arditti, Renate Duelli Klein, and Shelley Minden, 331–355. London: Pandora Press.

Huebner, Bryce. 2014. *Macrocognition: A Theory of Distributed Minds and Collective Intentionality*. New York: Oxford University Press.

Hull, David L., and Michael Ruse, eds. 1998. *The Philosophy of Biology*. New York: Oxford University Press.

Hypatia: A Journal of Feminist Philosophy 24 (4) (2009): 189–237.

Intemann, Kristin. 2010. "25 Years of Feminist Empiricism and Standpoint Theory: Where Are We Now?" *Hypatia* 25 (4): 778–796.

Intercontinental Cry. 2016. "Forced Sterilization of 272,000 Indigenous Women 'Not a Crime against Humanity' Public Prosecutor." September 13. https:// intercontinentalcry.org/forced-sterilization-272000-indigenous-women-not-crime -humanity-public-prosecutor/.

International Journal of Disability, Community, and Rehabilitation 12 (2) (2013). Special issue on "Eugenics," ed. Gregor Wolbring.

Jaggar, Alison. 1983. *Feminist Politics and Human Nature*. Totowa, NJ: Rowman and Littlefield.

Janis, Irving. 1972. *Victims of Groupthink: A Psychological Study of Foreign-Policy Decisions and Fiascoes*. Boston: Houghton Mifflin. Revised 2nd ed. titled *Groupthink: Psychological Studies of Policy Decisions and Fiascoes*. Boston: Houghton Mifflin, 1982.

Jankovic, Marija, and Kirk Ludwig, eds. In press. *Routledge Handbook on Collective Intentionality*. New York: Routledge.

Jarecki, Andrew. 2003. *Capturing the Friedmans*. HBO Documentary Films.

Jelliffe, Smith Ely, Elizabeth B. Muncey, and Charles B. Davenport. 1913. "Hunting-ton's Chorea: A Study in Heredity." *Journal of Nervous and Mental Disease* 40 (12): 796–799.

Johnson, Corey. 2013. "Female Inmates Sterilized in California Prisons without Approval." Center for Investigative Reporting. http://cironline.org/reports/female-inmates-sterilized-california-prisons-without-approval-4917.

Kaelber, Lutz. n.d. *Eugenics: Compulsory Sterilization in 50 American States*. http://www.uvm.edu/%7Elkaelber/eugenics/.

Kafer, Alison. 2013. *Feminist, Queer, Crip*. Bloomington: Indiana University Press.

Kahane, David, David Sharp, and Martin Tweedale. 1998. "Report of the MacEachran Subcommittee, Department of Philosophy, April 1998." University of Alberta, https://s3.amazonaws.com/bmcmahen/maceachran_report.pdf.

Keil, Frank C. 1989. *Concepts, Kinds, and Cognitive Development*. Cambridge, MA: MIT Press.

Kersten, Luke, and Robert A. Wilson. 2016. "The Sound of Music, Externalist Style." *American Philosophical Quarterly* 53 (2): 139–154.

Kersten, Luke. 2014. "Alberta Passes First Amendment to the Sexual Sterilization Act." Pathways module, EugenicsArchives.ca. http://eugenicsarchive.ca/discover/tree/5172e81ceed5c6000000001d.

Kersten, Luke. 2014. "Alberta Passes Sterilization Act." Pathways module, EugenicsArchives.ca. http://eugenicsarchive.ca/discover/timeline/5172e81ceed5c6000000001d.

Kevles, Daniel. [1985] 1995. *In the Name of Eugenics*. Cambridge, MA: Harvard University Press.

Kevles, Daniel, and Leroy Hood, eds. 1992. *The Code of Codes: Scientific and Social Issues in the Human Genome Project*. Cambridge, MA: Harvard University Press.

Kitcher, Philip. 1992. "The Naturalists Return." *Philosophical Review* 101:53–114.

Kitcher, Philip. 1996. *The Lives to Come: The Genetic Revolution and Human Possibilities*. New York: Simon and Schuster.

Kitcher, Philip, ed. 2003. "Utopian Eugenics and Social Inequality." In *In Mendel's Mirror: Philosophical Reflections on Biology*, 258–282. New York: Oxford University Press. Originally published in *Controlling Our Destinies: Historical, Philosophical, Ethical, and Theological Perspectives on the Human Genome Project*, ed. Phillip R. Sloan, 229–262. Notre Dame, IN: University of Notre Dame Press, 2000.

Kite, Elizabeth. 1912. "Two Brothers." *Survey* 27:1861–1864.

Kite, Elizabeth. 1913. "The Pineys." *Survey* 31:7–13.

Kittay, Eva Feder. 2010. "The Personal Is Philosophical Is Political: A Philosopher and Mother of a Cognitively Disabled Person Sends Notes from the Battlefield." In *Cognitive Disability and Its Challenge to Moral Philosophy*, ed. Eva Feder Kittay and

Lician Carlson, 393–413. New York: Wiley-Blackwell. Originally published in *Metaphilosophy* 40 (3–4) (2009): 606–627.

Kittay, Eva Feder, and Licia Carlson, eds. 2010. *Cognitive Disability and Its Challenge to Moral Philosophy*. New York: Wiley-Blackwell.

Kline, Wendy. 2005. *Building a Better Race: Gender, Sexuality, and Eugenics from the Turn of the Century to the Baby Boom*. Berkeley: University of California Press.

Kohlman, Michael. 2014. "Today and Tomorrow: *To-day and To-morrow* Book Series." Connections module, EugenicsArchives.ca. http://eugenicsarchive.ca/discover/connections/546d00a8dabeefbb1a000001.

Koop, C. Everett, and Francis A. Schaeffer. 1979. *Whatever Happened to the Human Race*. Westchester, IL: Crossway Books.

Kostir, Mary Storer. 1916. *The Family of Sam Sixty*. Mansfield: Press of Ohio State Reformatory.

Krueger, Joel. 2014. "Affordances and the Musically Extended Mind." *Frontiers in Psychology* 4 (1003): 1–9.

Landman, J. H. 1932. *Human Sterilization: The History of the Sexual Sterilization Movement*. New York: Macmillan.

Larson, Edward J. 1995. *Sex, Race, and Science: Eugenics in the Deep South*. Baltimore, MD: Johns Hopkins University Press.

Laughlin, Harry Hamilton. 1922. *Eugenical Sterilization in the United States*. Chicago: Psychopathic Laboratory of the Municipal Court of Chicago; https://archive.org/details/cu31924013882109.

Lenart, Bartlomiej A. 2013. "Shadow People: Relational Personhood, Extended Diachronic Personal Identity, and Our Moral Obligations toward Fragile Persons." Ph.D. dissertation, University of Alberta.

Leung, Colette. 2014. "Popular Culture." Encyclopedia module, EugenicsArchive.ca. http://eugenicsarchive.ca/discover/encyclopedia/535eed7a7095aa000000024a.

Leung, Colette. 2014. "Provincial Training School." Institutions module, EugenicsArchives.ca. http://eugenicsarchive.ca/discover/institutions/map/517da 50e9786fa0a73000001.

Le Vann, Leonard J. 1959. "Trifluoperazine Dihydrochloride: An Effective Tranquillizing Agent for Behavioural Abnormalities in Defective Children." *Canadian Medical Association Journal* 80 (2): 123–124.

Le Vann, Leonard J. 1961. "Thioridazine (Mellaril): A Psycho-sedative Virtually Free of Side-effects." *Alberta Medical Bulletin* 26:144–147.

Le Vann, Leonard J. 1968. "A New Butyrophenone: Trifluperidol: A Psychiatric Evaluation in a Pediatric Setting." *Canadian Psychiatric Association Journal* 13:271–273.

Le Vann, Leonard J. 1969. "Haloperidol in the Treatment of Behavioral Disorders in Children and Adolescents." *Canadian Psychiatric Association Journal* 14:217–220.

Le Vann, Leonard J. 1971. "Clinical Comparison of Haloperidol and Chlorpromazine in Mentally Retarded Children." *American Journal of Mental Deficiency* 75:719–723.

Levine, Philippa, and Alison Bashford. 2010. "Introduction: Eugenics and the Modern World." In *The Oxford Handbook of the History of Eugenics*, ed. Alison Bashford and Philippa Levine, 3–24. New York: Oxford University Press.

Lewontin, Richard. 1982. *Human Diversity*. New York: Scientific American Library.

Living Archives on Eugenics in Western Canada. 2008–2009. "Thinking in Action" series, Living Archives on Eugenics Blog. https://whatsortsofpeople.wordpress .com/2009/02/13/all-wrapped-up-complete-thinking-in-action-series/.

Living Archives on Eugenics in Western Canada. 2014. Institutions module, EugenicsArchives.ca. http://eugenicsarchive.ca/discover/institutions.

Living Archives on Eugenics in Western Canada. 2014. "Leilani Muir." Our Stories module, EugenicsArchives.ca. http://eugenicsarchive.ca/discover/our-stories/leilani.

Living Archives on Eugenics in Western Canada. 2014. "Margaret Thompson Panel Discussion." https://www.youtube.com/watch?v=jpAwWW4wu38.

Living Archives on Eugenics in Western Canada. 2014. "Nick Supina III." Interviews+ module, EugenicsArchives.ca. http://eugenicsarchive.ca/discover/interviews.

Living Archives on Eugenics in Western Canada. 2014. "Nick Supina III." Our Stories module, EugenicsArchives.ca. http://eugenicsarchive.ca/discover/our-stories/nick.

Living Archives on Eugenics in Western Canada. 2014. Our Stories module, Eugenics Archives.ca. http://eugenicsarchive.ca/discover/our-stories.

Living Archives on Eugenics in Western Canada. 2014. "Sandra Anderson." Interviews+ module, EugenicsArchives.ca. http://eugenicsarchive.ca/discover/interviews.

Living Archives on Eugenics in Western Canada. 2014. "Velvet Martin." Our Stories module, EugenicsArchives.ca. http://eugenicsarchive.ca/discover/our-stories/velvet.

Lloyd, G. E. R. 2012. *Being, Humanity, and Understanding*. New York: Oxford University Press.

Lloyd, Sharon A., and Susanne Sreedhar. 2014. "Hobbes's Moral and Political Philosophy." *Stanford Encyclopedia of Philosophy* (Spring edition), ed. Edward N. Zalta. https://plato.stanford.edu/archives/spr2014/entries/hobbes-moral/.

Lombardo, Paul. 2008. *Three Generations, No Imbeciles: Eugenics, the Supreme Court, and Buck v. Bell*. Baltimore, MD: Johns Hopkins University Press.

Lombardo, Paul, ed. 2011. *A Century of Eugenics in America: From the Indiana Experiment to the Human Genome Era*. Bloomington: Indiana University Press.

Lombardo, Paul, and Greg Dorr. 2000. "Eugenics Bibliography." http://buckvbell .com/othermaterial.html.

Longino, Helen. 1990. *Contextual Empiricism*. Princeton, NJ: Princeton University Press.

Longino, Helen. 2002. *The Fate of Knowledge*. Princeton, NJ: Princeton University Press.

Ludmerer, Kenneth M. 1972. *Genetics and American Society: A Historical Approach*. Baltimore, MD: Johns Hopkins University Press.

Lukács, Georg. [1923] 1971. *History and Class Consciousness: Studies in Marxist Dialectics*. Cambridge, MA: MIT Press. Originally published in German.

Lundborg, Herman. 1922. "The Swedish State-Institute of Race-Biology." *Acta Medica Scandinavica* 56 (1): 371–392.

MacEachran, John. 1932. "A Philosopher Looks at Mental Hygiene." *Mental Hygiene* 16:101–119.

MacEachran, John. 1932. "Crime and Punishment." Address to the United Farm Women's Association of Alberta. Reprinted in *The Press Bulletin* 17 (6): 1–4.

MacEachran, John. 1934. "Social Legislation in the Province of Alberta." Report to the Government of Alberta (author's copy).

MacIver, Malcolm. 2009. "Neuroethology: From Morphological Computation to Planning." In *The Cambridge Handbook of Situated Cognition*, ed. Philip Robbins and Murat Aydede, 480–504. New York: Cambridge University Press.

MacIver, Malcolm A., Lars Schmitz, Ugurcan Mugan, Todd D. Murphey, and Curtis D. Mobley. 2017. "Massive Increase in Visual Range Preceded the Origin of Terrestrial Vertebrates." *Proceedings of the National Academy of Sciences* 114 (12): E2375–E2384.

MacKenzie, Donald A. 1981. *Statistics in Britain, 1865–1930: The Social Construction of Scientific Knowledge*. Edinburgh: Edinburgh University Press.

Macpherson, C. B. 1962. *The Political Theory of Possessive Individualism*. Oxford: Clarendon Press.

Mahowald, Mary B. 1998. "A Feminist Standpoint." In *Disability, Difference, Discrimination: Perspectives on Justice in Bioethics and Public Policy*, ed. Anita Silvers, David T. Wasserman, and Mary B. Mahowald, 209–252. Lanham, MD: Rowman & Littlefield.

Maudsley, Henry. 1867. *The Physiology and Pathology of the Mind.* New York: D. Appleton and Company.

McBryde-Johnson, Harriet. 2003. "Unspeakable Conversations." *New York Times Magazine,* February 16.

McConvell, Patrick, and Helen Gardner. 2013. "The Descent of Morgan in Australia: Kinship Representation from the Australian Colonies." *Structure and Dynamics* 6 (1): 1–23. http://escholarship.org/uc/item/5711t341.

McConvell, Patrick, and Helen Gardner. 2015. *Southern Anthropology: A History of Fison and Howitt's Kamilaroi and Kurnai.* New York: Palgrave Macmillan.

McCulloch, Oscar. 1888. "The Tribe of Ishmael: A Study in Social Degradation." *Proceedings of the National Conference of Charities and Correction,* 154–159.

McLaren, Angus. 1990. *Our Own Master Race: Eugenics in Canada, 1885–1945.* Toronto: McClelland & Stewart.

McMahan, Jeff. 2010. "Cognitive Disability and Cognitive Enhancement." In *Cognitive Disability and Its Challenge to Moral Philosophy,* ed. Eva Feder Kittay and Lician Carlson, 345–367. New York: Wiley-Backwell. Originally published in *Metaphilosophy* 40 (3–4) (2009): 582–605.

Meade, Michelle L., Celia B. Harris, Penny Van Bergen, John Sutton, and Amanda J. Barnier, eds. 2017. *Collaborative Remembering: Theories, Research, and Applications.* New York: Oxford University Press.

Meltzoff, Andrew. 2007. "Like Me: A Foundation for Social Cognition." *Developmental Science* 19 (1): 126–134.

Meltzoff, Andrew N., and N. Keith Moore. 1977. "Imitation of Facial and Manual Gestures by Human Neonates." *Science* 198 (4312): 75–78.

Menary, Richard, ed. 2010. *The Extended Mind.* Cambridge, MA: MIT Press.

Mitchell, David, and Laura Snyder. 2003. "The Eugenic Atlantic: Race, Disability, and the Making of an International Eugenic Science, 1800–1945." *Disability and Society* 18 (7): 843–864.

Molnar, Stephen. 2006. *Human Variation: Races, Types, and Ethnic Groups,* 6th ed. New York: Routledge.

Moore, G. E. 1903. *Principia Ethica.* New York: Cambridge University Press.

Muir, Leilani. 2014. *A Whisper Past: Childless after Eugenic Sterilization in Alberta.* Victoria, BC: Friesen Press.

Muir v. Alberta. 1996. *Dominion Law Reports* 132 (4th series), 695–762, 1996 CanLII 7287 (QB AB).

Myerson, Abraham, James B. Ayer, Tracy J. Putnam, Clyde E. Keeler, and Leo Alexander. 1936. *Eugenical Sterilization—A Reorientation of the Problem. By the Committee of the American Neurological Association for the Investigation of Eugenical Sterilization*. New York: Macmillan.

Nathan, Debbie, and Michael Snedeker. [1995] 2001. *Satan's Silence: Ritual Abuse and the Making of a Modern American Witch Hunt*. Lincoln, NE: Author's Choice Press, an imprint of iUniverse.com, Inc.

Natoli, Jamie L., and Deborah L. Ackerman, Suzanne McDermott, and Janice G. Edwards. 2013. "Prenatal Diagnosis of Down Syndrome: A Systematic Review of Termination Rates (1995–2011)." *Prenatal Diagnosis* 32 (2): 142–153.

NeuroQueer. n.d. http://neuroqueer.blogspot.ca/.

Noë, Alva. 2004. *Action in Perception*. Cambridge, MA: MIT Press.

Nordau, Max. 1898. *Degeneration*. London: William Heineman.

Nye, Robert. 1975. *The Origins of Crowd Psychology: Gustav LeBon and the Crisis of Mass Democracy in the Third Republic*. Beverly Hills, CA: Sage Publications.

O'Brien, Gerald. 2013. *Framing the Moron: The Social Construction of Feeble-Mindedness in the American Eugenic Era*. Manchester: Manchester University Press.

Olick, Jeffrey K., Vered Vinitzky-Seroussi, and Daniel Levy, eds. 2011. *The Collective Memory Reader*. New York: Oxford University Press.

Oostenbroek, Janine, Thomas Suddendorf, Mark Nielsen, Jonathan Redshaw, Siobhan Kennedy-Costantini, Jacqueline Davis, Sally Clark, and Virginia Slaughter. 2016. "Comprehensive Longitudinal Study Challenges the Existence of Neonatal Imitation in Humans." *Current Biology* 26 (10): 1334–1338.

Osborn, Frederick. 1940. *Preface to Eugenics*. New York: Harper and Brothers.

Parens, Erik, and Adrienne Asch. 2000. "Disability Rights Critique of Prenatal Genetic Testing: Reflections and Recommendations." In *Prenatal Testing and Disability Rights*, ed. Erik Parens and Adrienne Asch, 3–43. Washington, DC: Georgetown University Press. Originally published in *Hastings Center Report* (Sept.–Oct 1999), S1–22.

Paul, Julius. 1965. *"… Three Generations of Imbeciles Are Enough …" State Eugenic Sterilization Laws in American Thought and Practice*. Manuscript. http://buckvbell .com/pdf/JPaulmss.pdf.

Paul, Laurie. 2015. *Transformative Experience*. New York: Oxford University Press.

Peace, William J., and Claire Roy. 2014. "Scrutinizing Ashley X: Presumed Medical 'Solutions' vs. Real Social Adaptation." *Journal of Philosophy, Science & Law* 14 (3) (July): 33–52.

Pearson, Karl. 1914. *The Life, Letters, and Labours of Francis Galton*, vol. 1. Cambridge: Cambridge University Press. Volumes 2 and 3 published in 1924 and 1930. http://galton.org/pearson/.

Pernick, Martin S. 1996. *The Black Stork: Eugenics and the Death of "Defective" Babies in American Medicine and Motion Pictures Since 1915*. New York: Oxford University Press.

Porter, Roy, and David Wright, eds. 2003. *The Confinement of the Insane: International Perspectives, 1800–1965*. Cambridge: Cambridge University Press.

Proctor, Robert N. 1988. *Racial Hygiene: Medicine under the Nazis*. Cambridge, MA: Harvard University Press.

Province of Alberta. 1928. *Sexual Sterilization Act of Alberta*. Statutes of the Province of Alberta, The Alberta Law Collection. http://www.ourfutureourpast.ca/law/page.aspx?id=2906151.

Province of Alberta. 1937. *An Act to Amend the Sexual Sterilization Act*. Statutes of the Province of Alberta, The Alberta Law Collection. http://www.ourfutureourpast.ca/law/page.aspx?id=2968369.

Province of Alberta. 1942. *An Act to Amend the Sexual Sterilization Act*. Statutes of the Province of Alberta, The Albert Law Collection. http://www.ourfutureourpast.ca/law/page.aspx?id=2914363.

Puar, Jasbir. 2013. "The Cost of Getting Better: Ability and Debility." In *The Disability Studies Reader*, 4th ed., ed. Lennard Davis, 177–184. New York: Routledge.

Rabinowitz, Dorothy. 2003. *No Crueler Tyrannies: Accusation, False Witness, and Other Terrors of Our Times*. New York: Free Press.

Rafter, Nicole. 1988. *White Trash: The Eugenic Family Studies 1977–1919*. Boston: Northeastern University Press.

Reiss, Julian, and Jan Sprenger. 2014. "Scientific Objectivity." *Stanford Encyclopedia of Philosophy*. https://plato.stanford.edu/entries/scientific-objectivity/.

Ridge, Michael. 2014. "Moral Non-naturalism." *Stanford Encyclopedia of Philosophy*. https://plato.stanford.edu/entries/moral-non-naturalism/.

Rimke, H., and A. Hunt. 2001. "From Sinners to Degenerates: The Medicalization of Morality in the 19th Century." *History of the Human Sciences* 15 (1): 59–88.

Rivers, W. H. R. [1910] 1968. "The Genealogical Method of Anthropological Inquiry." In *Kinship and Social Organization*, 97–112. New York: The Althone Press.

Rogers, Arthur, and Maud Merrill. 1919. *Dwellers in the Vale of Siddum*. Boston: Richard G. Badger.

Roper, Allen G. 1913. *Ancient Eugenics*. London: Clivenden Press.

Rose, Nikolas. 1985. *The Psychological Complex: Psychology, Politics and Society in England 1869–1939*. London: Routledge and Kegan Paul.

Rose, Nikolas. 2007. *The Politics of Life Itself: Biomedicine, Power, and Subjectivity in the Twenty-First Century*. Princeton, NJ: Princeton University Press.

Rose, Nikolas, and Joelle M. Abi-Rached. 2012. *Neuro: The New Brain Sciences and the Management of the Mind*. Princeton, NJ: Princeton University Press.

Roskies, Ethel. 1972. *Abnormality and Normality: The Mothering of Thalidomide Children*. Ithaca, NY: Cornell University Press.

Rouse, Joseph. 1996. "Feminism and the Social Construction of Scientific Knowledge." In *Feminism, Science and the Philosophy of Science*, ed. Lynne Hankinson Nelson and James Nelson, 195–215. Dordrecht: Kluwer Academic Publishers.

Rupert, Robert. 2009. *Cognitive Systems and the Extended Mind*. New York: Oxford University Press.

Rysiew, Patrick. 2016. "Epistemic Contextualism." *Stanford Encyclopedia of Philosophy* (Winter edition), ed. Edward N. Zalta. https://plato.stanford.edu/archives/win2016/entries/contextualism-epistemology/.

Samson, Amy. 2014. "Eugenics in the Community: Gendered Professions and Eugenic Sterilization in Alberta, 1928–1972." *Canadian Bulletin of Medical History* 31 (1): 143–164.

Samson, Amy. 2014. "Eugenics in the Community: The United Farm Women of Alberta, Public Health Nursing, Teaching, Social Work, and Sexual Sterilization in Alberta, 1928–1972." PhD dissertation, University of Saskatchewan.

Samson, Amy. 2014. "Guidance Clinics." Encyclopedia module, EugenicsArchive.ca. http://eugenicsarchive.ca/discover/encyclopedia/555427c835ae9d9e7f000030.

Savulescu, Julian. 2001. "Procreative Beneficence: Why We Should Select the Best Children." *Bioethics* 15 (5/6): 413–426.

Savulescu, Julian. 2008. "Procreative Beneficence: Reasons to Not Have Disabled Children." In *The Sorting Society: The Ethics of Genetic Screening and Therapy*, ed. Loane Skene and Janna Thompson, 51–68. New York: Cambridge University Press.

Savulescu, Julian, and Guy Kahane. 2009. "The Moral Obligation to Create Children with the Best Chance of the Best Life." *Bioethics* 23 (5): 274–290.

Saxton, Marsha. 1984. "Born and Unborn: The Implications of Reproductive Technologies for People with Disabilities." In *Test-Tube Women: What Future for Motherhood*, ed. Rita Arditti, Renate Duelli Klein, and Shelley Minden, 298–312. London: Pandora Press.

Saxton, Marsha. 1997. "Disability Rights and Selective Abortion." In *In Abortion Wars: A Half Century of Struggle, 1950–2000*, ed. Rickie Solinger, 374–395. Berkeley: University of California Press.

Saxton, Marsha. 2000. "Why Members of the Disability Community Oppose Prenatal Diagnosis and Selective Abortion." In *Prenatal Testing and Disability Rights*, ed. Erik Parens and Adrienne Asch, 147–164. Washington, DC: Georgetown University Press.

Scandinavian Journal of History 24 (2) (1999). Special issue on "Eugenics in Scandinavia," ed. Gunnar Broberg and Mathias Tyden.

Schroeder, David A., and William G. Grazziano, eds. 2015. *The Oxford Handbook of Prosociality*. New York: Oxford University Press.

Schuster, Edgar. 1912. *Eugenics*. London: Collins Clear Type Press.

Scull, Andrew. 1979. *Museums of Madness: The Social Organization of Insanity in Nineteenth-Century England*. New York: St. Martin's Press.

Searle, John R. 1995. *The Construction of Social Reality*. New York: Free Press

Searle, John R. 2010. *Making the Social World: The Structure of Human Civilization*. New York: Oxford University Press.

Sequenzia, Amy. n.d. *Non-speaking Autistic Speaking*. http://nonspeakingautisticspeaking.blogspot.ca/.

Sessions, Mina A. 1918. *The Feeble-Minded in a Rural County of Ohio*. Bulletin number 6 of the Bureau of Juvenile Research. Mansfield: Ohio State Reformatory Printers.

Shakespeare, Tom. 2003. *Disability Rights and Wrongs Revisited*. New York: Routledge.

Shakespeare, Tom. 2005. "The Social Context of Individual Choice." In *Quality of Life and Human Difference*, ed. David Wasserman, Jerome Bickenbach, and Robert Wachbroit, 217–236. New York: Cambridge University Press.

Shotwell, Alexis. 2011. *Knowing Otherwise: Race, Gender, and Implicit Understanding*. University Park: Pennsylvania State University Press.

Signs: Journal of Women and Culture 22 (2) (1997). Special issue.

Silvers, Anita. 1998. "A Fatal Attraction to Normalizing." In *Enhancing Human Traits: Ethical and Social Implications*, ed. Erik Parens, 95–121. Washington, DC: Georgetown University Press.

Silvers, Anita. 2016. "Disability and Normality." In *Routledge Companion to Philosophy of Medicine*, ed. Miriam Solomon, Jeremy R. Simon, and Harold Kincaid, 36–47. New York: Routledge.

Singer, Peter. 2010. "Speciesism and Moral Status." In *Cognitive Disability and Its Challenge to Moral Philosophy*, ed. Eva Feder Kittay and Lician Carlson, 331–343. New York: Wiley-Backwell. Originally published in *Metaphilosophy* 40 (3–4) (2009): 567–581.

Skotko, Brian G. 2009. "With New Prenatal Testing, Will Babies with Down Syndrome Slowly Disappear?" *Archives of Disease in Childhood* 94:823–826.

Smith, David Livingstone. 2011. *Less Than Human: Why We Demean, Enslave, and Exterminate Others*. New York: St. Martin's Press.

Smith, David Livingstone. 2014. "Dehumanization, Essentialism, and Moral Psychology." *Philosophy Compass* 9:814–824.

Smith, David Livingstone. 2014. "Dehumanization: Psychological Aspects." Encyclopedia module, EugenicsArchives.ca. http://eugenicsarchive.ca/discover/encyclopedia/53d82f9d4c879d000000000a.

Smith, Dorothy. 1974. "Women's Perspective as a Radical Critique of Sociology." *Sociological Inquiry* 44:7–13.

Sneddon, Andrew. 2011. *Like-Minded: Externalism and Moral Psychology*. Cambridge, MA: MIT Press.

Society of Obstetricians and Gynaecologists of Canada. 2007. "Prenatal Screening for Fetal Aneuploidy." *Journal of Obstetricians and Gynaecologists of Canada* 187 (February): 146–161.

Society of Obstetricians and Gynaecologists of Canada. 2012. "Counselling Considerations for Prenatal Genetic Screening." *Journal of Obstetricians and Gynaecologists of Canada* 277 (May):489–493.

Solomon, Miriam. 2001. *Social Empiricism*. Cambridge, MA: MIT Press.

Sparrow, Rob. 2014. "In Vitro Eugenics." *Journal of Medical Ethics* 40 (11): 725–731.

Sperber, Dan. 1996. *Explaining Culture: A Naturalistic Approach*. New York: Cambridge University Press.

Stahnisch, Frank W., and Erna Kurbegović, eds. In press. *Exploring the Relationship of Eugenics and Psychiatry: Canadian and Trans-Atlantic Perspectives 1905–1972*. Edmonton, AB: Athabasca University Press.

Steeup, Matthias. 2016. "Epistemology." *Stanford Encyclopedia of Philosophy* (Fall edition), ed. Edward N. Zalta. https://plato.stanford.edu/archives/fall2016/entries/epistemology/.

Stephens, T., and R. Brynner. 2001. *The Dark Remedy: The Impact of Thalidomide*. New York: Perseus Publishing.

The Sterilization of Leilani Muir. 1996. Dir. Glynis Whiting. Ottawa: National Film Board of Canada.

Stern, Alexandra Minna. 2005. *Eugenic Nation: Faults and Frontiers of Better Breeding in Modern America.* Berkeley: University of California Press

Stern, Alexandra Minna. 2014. "Marriage." Encyclopedia module, EugenicsArchives. ca. http://eugenicsarchive.ca/discover/encyclopedia/535eeccb7095aa000000023b.

Stoddard, Lothrop. 1920. *The Rising Tide of Color against White World-Supremacy.* New York: Charles Scribner's Sons.

Surviving Eugenics. 2015. Dir. Jordan Miller, Nicola Fairbrother, and Robert A. Wilson. Vancouver, BC: Moving Images Distribution.

Taylor, Ashley. 2015. "Expressions of 'Lives Worth Living' and Their Foreclosure through Philosophical Theorizing on Moral Status and Intellectual Disability." In *Foucault and the Government of Disability*, rev. ed., ed. Shelley Tremain, 372–395. Ann Arbor: University of Michigan Press.

Thomson, Mathew. 1998. *The Problem of Mental Deficiency: Eugenics, Democracy, and Social Policy in Britain 1870–1959.* Oxford: Oxford University Press.

Thomson, Mathew. 2010. "Disability, Psychiatry, and Eugenics." In *The Oxford Handbook of the History of Eugenics*, ed. Alison Bashford and Philippa Levine, 116–131. New York: Oxford University Press.

Tomasello, Michael. 2014. *A Natural History of Human Thinking.* Cambridge, MA: Harvard University Press.

Tredgold, Arthur. 1910. "The Feeble-Minded." *Contemporary Review* 97 (534): 717–727.

Tremain, Shelley. 2001. "On the Government of Disability." *Social Theory and Practice* 27 (4): 617–636.

Tremain, Shelley, ed. [2005] 2015. *Foucault and the Government of Disability.* Ann Arbor: University of Michigan Press.

Trent, James W. Jr. 1994. *Inventing the Feeble Minded: A History of Mental Retardation in the United States.* Berkeley: University of California Press.

Truth and Reconciliation Commission of Canada and National Centre for Truth and Reconciliation. n.d. www.trc.ca.

Tuck, Richard. 1989. *Hobbes.* Oxford: Oxford University Press.

Tuomela, Raimo. 2007. *The Philosophy of Sociality: The Shared Point of View.* New York: Oxford University Press.

Turda, Marius. 2010. *Modernism and Eugenics.* Basingstoke: Palgrave.

Turda, Marius. 2010. "Race, Science, and Eugenics in the Twentieth-Century." In *The Oxford Handbook of the History of Eugenics*, ed. Alison Bashford and Philippa Levine, 62–79. New York: Oxford University Press.

UNICEF Convention on the Rights of the Child. 1989. http://digitalcommons .ilr.cornell.edu/cgi/viewcontent.cgi?article=1007&context=child.

van Ginneken, Jaap. 1992. *Crowds, Psychology, and Politics 1871–1899*. New York: Cambridge University Press.

Wahlsten, Douglas. 1997. "Leilani Muir versus the Philosopher King: Eugenics on Trial in Alberta." *Genetica* 99:185–198.

Wasserman, David, Adrienne Asch, Jeffrey Blustein, and Daniel Putnam. 2013. "Disability and Justice." *Stanford Encyclopedia of Philosophy*. https://plato.stanford.edu/ entries/disability-justice/.

Watkins, J. W. N. 1965. *Hobbes's System of Ideas*. London: Hutchison.

Watson, James. 2001. *A Passion for DNA: Genes, Genomes, and Society*. Cold Spring Harbor, NY: Cold Spring Harbor Laboratory Press.

Weber, Barbara. 2013. *Zwischen Vernunft und Mitgefuhl*. Freiburg: Alber Publishers.

Weindling, Paul. 1989. *Health, Race and German Politics between National Unification and Nazism, 1870–1945*. Cambridge: Cambridge University Press.

Weindling, Paul. 2010. "German Eugenics and the Wider World: Beyond the Racial State." In *The Oxford Handbook of the History of Eugenics*, ed. Alison Bashford and Philippa Levine, 315–331. New York: Oxford University Press.

Weinraub, Bernard. 1981. "Reagan Nominee for Surgeon General Runs into Obstacles on Capitol Hill." *New York Times*, April 7, A16/6.

Wells, Rob. 2010. "In the Matter of a Petition for the Removal of a Member of the Order of Canada" (author's copy).

Wendell, Susan. 1996. *The Rejected Body: Feminist Philosophical Reflections on Disability*. New York: Routledge.

Wilfond, Benjamin S., Paul Steven Miller, Carolyn Korfiatis, Douglas S. Diekema, Denise M. Dudzinski, and Sara Goering. 2010. "Navigating Growth Attenuation in Children with Profound Disabilities." *Hastings Center Report* 40 (6): 27–40.

Will, George F. 2007. "Golly, What *Did* Jon Do?" *Newsweek*, January 29.

Wilson, Edward O. 1975. *Sociobiology: The New Synthesis*. Cambridge, MA: Belknap Press.

Wilson, Robert A. 1995. *Cartesian Psychology and Physical Minds: Individualism and the Sciences of the Mind*. New York: Cambridge University Press.

Wilson, Robert A. 2004. *Boundaries of the Mind: The Individual in the Fragile Sciences: Cognition.* New York: Cambridge University Press.

Wilson, Robert A. 2007. "Social Reality and Institutional Facts: Sociality within and without Intentionality." In *Intentional Acts and Institutional Facts: Essays on John Searle's Social Ontology*, ed. Savas L. Tsohatzidis, 139–153. Dordrecht: Springer.

Wilson, Robert A. 2010. "Extended Vision." In *Perception, Action and Consciousness: Sensorimotor Dynamics and Two Visual Systems*, ed. Nivedita Gangopadhyay, Michael Madary, and Finn Spicer, 277–290. New York: Oxford University Press.

Wilson, Robert A. 2014. "Eugenic Family Studies." Encyclopedia module, EugenicsArchives.ca. http://www.eugenicsarchives.ca/discover/encyclopedia/535ee bbb7095aa0000000225.

Wilson, Robert A. 2014. "Sorts of People." April 29. Encyclopedia module, EugenicsArchive.ca. http://eugenicsarchive.ca/discover/encyclopedia/535eee527095 aa000000025c.

Wilson, Robert A. 2014. "Ten Questions Concerning Extended Cognition." *Philosophical Psychology* 27 (1): 19–33.

Wilson, Robert A. 2015. "The Role of Oral History in Surviving a Eugenic Past." In *Beyond Testimony and Trauma: Oral History in the Aftermath of Mass Violence*, ed. Steven High, 119–138. Vancouver: University of British Columbia Press.

Wilson, Robert A. 2017. "Collective Intentionality in Non-human Animals." In *Routledge Handbook on Collective Intentionality*, ed. Marija Jankovic and Kirk Ludwig. New York: Routledge.

Wilson, Robert A. 2017. "Eugenics and Philosophy." *Oxford Bibliographies Online.* www.oxfordbibliographiesonline.com.

Wilson, Robert A. 2017. "Group-level Cognizing, Collaborative Remembering, and Individuals." In *Collaborative Remembering: How Remembering with Others Influences Memory*, ed. Michelle Meade, Amanda Barnier, Penny Van Bergen, Celia Harris, and John Sutton. New York: Oxford University Press.

Wilson, Robert A. Unpublished manuscript. *Relative Beings.*

Wilson, Robert A., and Andy Clark. 2009. "How to Situate Cognition: Letting Nature Take Its Course." In *The Cambridge Handbook of Situated Cognition*, ed. Philip Robbins and Murat Aydede, 55–77. New York: Cambridge University Press.

Wilson, Robert A., and Lucia Foglia. 2016. "Embodied Cognition." *Stanford Encyclopedia of Philosophy* (Winter edition), ed. Edward N. Zalta. https://plato.stanford.edu/archives/win2016/entries/embodied-cognition/.

Winship, Albert E. 1900. *Jukes-Edwards: A Study in Education and Heredity.* Harrisburg, PA: R. L. Meyers.

Witch Hunt. 2009. Dir. Don Hardy Jr. and Dana Nachman. Glendale, CA: Hard-Nac Movies, Your Half Media Group. www.witchhuntmovie.com.

Wolbring, Gregor. 2004. "Parents without Prejudice." In *Reflections from a Different Journey: What Adults with Disabilities Wish All Parents Knew*, ed. Stanley Klein and John Kemp, 18–22. New York: McGraw Hill.

Women With Disabilities Australia. 2013. "Dehumanised: The Forced Sterilisation of Women and Girls with Disabilities in Australia." WWDA Submission to the Senate Inquiry into the Involuntary or Coerced Sterilisation of People with Disabilities in Australia, March. http://wwda.org.au/papers/subs/subs2011/.

World Health Organization. [1980] 1993. *International Classification of Impairments, Disabilities, and Handicaps.* Geneva: World Health Organization. http://apps.who.int/iris/bitstream/10665/41003/1/9241541261_eng.pdf.

Wright, Suzanne Wright. 2013. "Autism Speaks to Washington—A Call for Action." http://www.autismspeaks.org/news/news-item/autism-speaks-washington-call-action.

Wylie, Alison. 2004. "Why Standpoint Matters." In *The Feminist Standpoint Theory Reader*, ed. Sandra Harding, 339–351. New York: Routledge. Originally published in *Science and Other Cultures: Issues in Philosophies of Science and Technology*, ed. Robert Figueroa and Sandra Harding, 26–48. New York: Routledge, 2003.

Wylie, Alison. 2012. "Feminist Philosophy of Science: Standpoint Matters." *Proceedings and Addresses of the American Philosophical Association* 86 (2): 47–76.

Yergeau, Melanie. n.d. *Aspie Rhetor.* http://aspierhetor.com/.

Young, Allan. 1995. *The Harmony of Illusions: Inventing Post-Traumatic Stress Disorder.* Princeton, NJ: Princeton University Press.

Zagorin, Peter. 2009. *Hobbes and the Law of Nature.* Princeton, NJ: Princeton University Press.

Index

23